"Organized around the seven themes of Catholic social teaching, this fine work makes clear how 'liturgy provides the necessary basis for Christian mission and social outreach,' since both liturgy and Catholic social teaching engage the entire person and relate to our entire life. In clear and accessible language, the authors highlight the link between corporate worship and human dignity. The book's discussion questions and prayer resources make it an ideal resource for pastoral teams and parish social ministers. Highly recommended!"

R. Scott Hurd
Vice President for Leadership Development
and Catholic Identity at Catholic Charities USA
Author of *Forgiveness: A Catholic Approach* and other works

"*Liturgy and Catholic Social Teaching* unites theologically rich and pastorally relevant reflection on the integral relationship between worship and everyday life. All who long to make more vital connections between the Eucharistic table and the work for justice and reconciliation around the tables of the world will find comfort, challenge, and hope in these pages."

Antonio Alonso
Composer, Author, and Theologian

"As the climate crisis unfolds, I am more convinced than ever that a radical change in the sphere of the heart—reignited by the recognition that all of creation is a sacrament—may be the only way to save ourselves from our rush towards the abyss. This book is a vital contribution and a call to action to this effort. It reminds us that our at the center of our liturgical practices is an encounter with the Creator and a celebration 'on the altar of the world' (*Laudato si'*, 236)."

Dan Misleh
Executive Director, Catholic Climate Covenant

"This book is an excellent summary of the seven themes of Catholic social teaching with each author incorporating texts from theologians, recent papal documents and statements that clarify and elucidate the topics. It will provide a rich resource for teachers, preachers and parish staff as they deepen their own understanding of CST and offer educational opportunities to others. The linkages between the themes and our beautiful liturgical cycle and rituals flow effortlessly and open vistas for personal and communal prayer. I used it with the RCIA group and know that the suggested penitential services in the appendix will be appreciated by pastors and parishioners alike for connecting values, actions, and Sacred Scripture. Each of the contributing authors has impressive credentials and provides a solid foundation to the discussion of very serious and timely topics."

Patricia M. Brown, SSMN
Retired Director of Catholic Charities of South Georgia
Diocese of Savannah

"*Liturgy and Catholic Social Teaching* is a valuable contribution to the pastoral challenge of a making Catholic social teaching a constitutive part of parish life. It is an important resource for parish social justice committees and worship committees to study together. Readers will come away with a greater understanding of the Church's rich tradition of social teaching and a deeper appreaction of the Eucharist as the source and summit of the Christian life. The appendix also includes excellent liturgical resources that parishes can use."

Tom Cordaro
Justice and Outreach Minister
St. Margaret Mary Parish, Naperville, IL

"This resource is an excellent attempt to examine the relationship between liturgy and social justice. I have not seen many books like this. It is truly pioneering. . . . The questions offered at the end of each chapter are provoking. The penitential services are inspiring and devotional. They offer so much variety, whether hymns or responsorial psalms, readings or profound reflective questions. The homily points too are very substantial, pregnant with ideas that can be developed. This book can be a useful resource manual for parishes, parishioners, and liturgists."

Fr. John D'Mello, PHD
Moral Theologian
Parochial Vicar, St. Patrick Church, Palm Beach Gardens, FL

The Liturgy and Catholic Social Teaching

Participation in Worship and the World

Kevin Ahern

Larry Dowling

Bernard Evans

Anne Y. Koester

Thomas Massaro, SJ

Dawn M. Nothwehr, OSF

Timothy P. O'Malley

Thomas Scirghi, SJ

Kate Ward

Nihil Obstat
Rev. Mr. Daniel G. Welter, JD
Chancellor
Archdiocese of Chicago
June 10, 2019

Imprimatur
Most Rev. Ronald A. Hicks
Vicar General
Archdiocese of Chicago
June 10, 2019

The *Nihil Obstat* and *Imprimatur* are declarations that the material is free from doctrinal or moral error, and thus is granted permission to publish in accordance with c. 827. No legal responsibility is assumed by the grant of this permission. No implication is contained herein that those who have granted the *Nihil Obstat* and *Imprimatur* agree with the content, opinions, or statements expressed.

This book was edited by Danielle A. Noe. Lauren L. Murphy was the production editor, and Kari Nicholls was the designer and production artist.

Cover art by Tracy Walker © LTP.

23 22 21 20 19   1 2 3 4 5

Printed in the United States of America

Library of Congress Control Number: 2019937419

ISBN 978-1-61671-509-0

LCST

From the Sunday Mass there flows a tide of charity destined to spread into the whole life of the faithful, beginning by inspiring the very way in which they live the rest of Sunday. . . .

Lived in this way, not only the Sunday eucharist but the whole of Sunday becomes a great school of charity, justice and peace. The presence of the Risen Lord in the midst of his people becomes an undertaking of solidarity, a compelling force for inner renewal, an inspiration to change the structures of sin in which individuals, communities and at times entire peoples are entangled. Far from being an escape, the Christian Sunday is a prophecy inscribed on time itself, a prophecy obliging the faithful to follow in the footsteps of the One who came "to preach good news to the poor, to proclaim release to captives and new sight to the blind, to set at liberty those who are oppressed, and to proclaim the acceptable year of the Lord" (Luke 4:18–19). In the Sunday commemoration of Easter, believers learn from Christ, and remembering his promise: "I leave you peace, my peace I give you" (John 14:27), they become in their turn *builders of peace*.

—*Dies Domini*, 72 and 73

CONTENTS

INTRODUCTION
Liturgy and Catholic Social Teaching

Bernard Evans

THE BIBLE TELLS US THAT WE MUST WORSHIP GOD AND LIVE OUR FAITH in every area of our lives. Catholics have two remarkable treasures to guide them in this lifelong and glorious task—the liturgy and Catholic social teaching. This book explores the relationship between the two. It begins with this introduction offering a brief account of what we mean by Catholic social teaching, followed by a historical survey of the development of these teachings from their beginning in the nineteenth century to the present moment. The survey includes a discussion of two complementary movements within the Church over the past one hundred years. One is a refocus on social justice, and the other is liturgical renewal. This is followed by a discussion of the points of contact between liturgy and Catholic social teachings.

What Is Catholic Social Teaching?

Liturgy engages the entire person and relates to our entire life, and so does Catholic social teaching. It is in that parallel that we find the strongest connection between public worship and daily actions in our communal lives.

Catholic social teaching guides us in how we are to live the Gospel in today's world beyond the more personal dimensions of daily living. These teachings show us what it means to be a disciple of Jesus Christ while facing war, poverty, racism, or any other challenges of modern times. They indicate how a Christian is to live in his or her context amid the challenges posed by social, economic, and political forces of any time and place. They help us discern how we are to live the Christian life here and now.

These social teachings of the Church do not provide black-and-white or detailed solutions to complex problems of the day. Rather, they offer the principles and the moral framework to lead us as we engage such challenges. In a sense, they furnish the connecting tissue between the Gospel accounts written in the first century and the ever-changing news headlines of the present day.

Catholic social teaching, for example, does not tell us "what would Jesus do in the face of climate change." It does suggest how his followers today should respond to this modern peril in light of what Jesus taught about love of neighbor and being in tune with all of creation. In short, these teachings draw on the Scriptures and the rest of the Catholic theological tradition to formulate moral direction relevant to living the Christian life in modern centuries.

Historical Survey of Catholic Social Teaching

A more complete analysis of the origins and development of Catholic social teaching would consider in greater detail the unfolding of Christian social ethics, especially Catholic moral theology, over the past two thousand years. The purpose of this introduction, however, requires a more condensed presentation of the biblical and theological foundations for these teachings as well as their historical beginning in the late nineteenth century. This survey concludes with a brief discussion of the modern liturgical renewal and the Catholic refocus on social justice, both of which impact the relationship between liturgy and Catholic social teachings.

The Scriptures provide the basic foundation for Catholic social teachings. In these sacred texts we find the rationale for why the Church and all Christians need to be engaged in the task of building a more just and caring world.

The prophets of the Old Testament noted that the peoples' unfaithfulness to God was seen in their refusal to respond to persons among them who were poor and marginalized. These were represented especially by the widows, orphans, and strangers. Caring for the marginalized and the powerless was a criterion for recognizing whether the Hebrew people were living in just relationships—with their neighbors and with their God. Isaiah[1] and Amos[2] even insisted that the peoples' fasting and worship were useless if these religious practices did not lead to just actions toward their neighbors.

1. See Isaiah 58:6–9.
2. See Amos 5:21–24.

The New Testament also emphasized that Christians must respond to the needs of persons living in poverty. At the beginning of his public ministry, Jesus made it clear that his own work included proclaiming Good News to the poor, announcing release to captives, and restoring sight to the blind while granting freedom to the oppressed.[3] Likewise, in the parable of the Last Judgment, Jesus warned that we will be judged by how well we respond to persons lacking food, water, and medical care. As long as you did it for the least of my sisters and brothers you did it for me—enter into the kingdom.[4]

Already in the first few centuries after his death, the early Christians were asking what it means to be a disciple of Jesus Christ in their societies. These theologians wrote about poverty and wealth and the Christian's responsibility toward people with greater needs than their own. Early Christian writers, like St. Ambrose of Milan and St. John Chrysostom, insisted that the goods of the earth are intended to meet the needs of all people. A Christian's relationship with God, they reiterated, is tied to one's relationship with other people, especially people who lack what is required for a basic, dignified life. Failure to meet their needs is a failure to meet the expectations that flow from our relationship with God.

From the earliest centuries through the Middle Ages to modern times, the Church keeps asking what it means to be a follower of Jesus Christ in any place and at any time. This inquiry always has concluded that to be a Christian places certain demands on our personal lives and on the social dimensions of those lives.

Those social contexts within which our lives unfold are always changing and presenting new challenges, questions, and issues to sort out. As the early Christians struggled with such matters as poverty and riches, so Christians in later centuries had to face these and other contentious topics of their day. Usury and slavery are further examples of difficult matters Christians, and especially Church leaders, addressed over many centuries. In more recent times the entire Church has focused on the morality of war, on capital punishment, and on caring for the environment.

In the late nineteenth century, Church leaders began to write about these moral teachings related to economic, political, and social issues, and

3. See Luke 4:18.
4. See Matthew 25:31–46.

they did so in a systematic way. These writings are what we now refer to as Catholic social teaching, and they represent the official teachings of the Church on these matters. This body of official teachings does not exhaust the entirety of the Catholic social tradition. The latter incudes writings, practices, and discussions by all members of the Church—lay and clerical—regarding social justice. Catholic social teaching, as used here, refers to the encyclicals and other documents issued by popes, councils, and synods.

The first of these documents was an encyclical in 1891 by Pope Leo XIII, *Rerum novarum* (*On the Condition of Labor*). This encyclical is an excellent example of Catholic social teaching. It responds to what at that time were major economic and social problems facing Europe and North America: the rights and responsibilities of workers, private ownership of property, and the looming threat of Marxist socialism. *Rerum novarum* examined each of these topics as they were experienced in the late nineteenth century and presented moral guidance for Catholic clergy and laity in dealing with these difficult modern challenges.

Since that time, the Vatican has given us occasional documents that constitute the body of official Catholic social teaching. The documents most commonly come in the form of papal encyclicals, but they do include other formats as well, for example, the Second Vatican Council's *Gaudium et spes* (*Pastoral Constitution on the Church in the Modern World*) in 1965, Pope Paul VI's 1971 apostolic letter *Octogesima adveniens* (*A Call to Action*) and the 1971 World Synod of Bishops' document, *Justitia in mundo* (*Justice in the World*). The number of Catholic social documents is not large—twelve to fifteen depending on what we choose to include among them.[5]

Again, each of these documents draws a connection between our Christian faith and the many challenges facing contemporary society. The documents remind us that our faith must express itself in our public, communal life. They offer basic Christian perspectives regarding the human person, and they remind us that we must organize society in ways that protect and honor the dignity of every person.

---

5. A helpful collection of these teachings can be found in the publication by David O'Brien and Thomas Shannon, *Catholic Social Thought: Encyclicals and Documents from Pope Leo XIII to Pope Francis*, 3rd rev. ed. (Maryknoll, NY: Orbis Books, 2016). This collection includes Pope Francis' beautiful encyclical, *Laudato si'* (*On Caring for Our Common Home*), the first encyclical to address contemporary concerns about caring for the earth ("our common home") and specifically to speak of climate change (see theme 7 on page 158).

Any discussion of liturgy and Catholic social teaching must take note of two parallel movements unfolding within the Church throughout the twentieth century. One is the refocus on social justice through teachings and practices; the other is the liturgical renewal.

As Catholic social teaching developed in the twentieth century there was a heightened emphasis on social justice within the Church. Examples from the Vatican and from the United States Catholic bishops illustrate this point.

In 1919, the United States Catholic bishops issued a pastoral letter calling for economic and social changes to benefit workers. Twenty years before the New Deal reforms of the Roosevelt administration, the bishops' *Program of Social Reconstruction* called for sweeping reforms to labor laws and practices. These included minimum wage legislation, unemployment insurance, child labor laws, the legal enforcement of workers' rights to organize, public housing for the working classes, and participation of workers in management and ownership of enterprises. The bishops also called for—in 1919—equal pay for women doing the same work as men.

Sixty-seven years later the United States Catholic bishops issued another pastoral letter, *Economic Justice for All*, which included a call for social justice and a clear statement of what that means. Social justice, they wrote, carries two related points. It means that every person has a duty to contribute to the well-being of society, and that society has a duty to enable persons to participate in this way. Thus, if individuals or groups suffer from poverty, discrimination, or homelessness, society must enact systemic changes to address these needs and to allow such persons to participate in their communities and contribute to the common good.[6]

This renewed emphasis on social justice is seen in Vatican documents during this century as well. In 1971, Pope Paul VI issued his apostolic letter, *Octogesima adveniens* (*A Call to Action*). This is the Catholic social document that first refers to the preferential option for the poor.[7] It also calls on the laity "without waiting passively for orders and directives, to take the initiative freely and to infuse a Christian spirit into the mentalities, customs, laws and structures of the community in which they live."[8]

6. See *Economic Justice for All*, 71.

7. See *Octogesima adveniens* (OA), 23; although the exact phrasing does not appear until the 1987 document, *Sollicitudo rei socialis* (see 42; and theme 4 in this book on page 128).

8. See OA, 48.

During that same year, the World Synod of Bishops gave us *Justitia in mundo* (*Justice in the World*). This social statement offered what in 1971 was a remarkable claim: "Action on behalf of justice and participation in the transformation of the world fully appear to us as a constitutive dimension of the preaching of the Gospel, or, in other words, of the Church's mission for the redemption of the human race and its liberation from every oppressive situation."[9] *Justitia in mundo* emphasized that every person has a right to be involved in decision-making processes that affect their lives—economic, social and political.[10] It reminded us as well that Christian love of neighbor and justice cannot be separated because "love implies an absolute demand for justice."[11]

Accompanying these teachings on social justice were various actions that demonstrated the Church's resolve to put justice theory into practice. In 1967, the Vatican, following the recommendation of the Second Vatican Council two years earlier, established the Pontifical Commission for Justice and Peace. The purpose of this commission (later renamed the Pontifical Council for Justice and Peace) would be to "encourage the catholic community to promote the progress in areas which are in want and foster social justice between nations."[12]

Two years later the United States Catholic bishops launched the Catholic Campaign for Human Development (CCHD). This is a justice education and grant-making program that provides millions of dollars each year to self-help, social change projects in low-income communities throughout the United States. These projects support the efforts of people to design their own local approaches to changing social structures and policies that undermine life and dignity, especially for persons who are economically poor or politically powerless. CCHD's justice education work helps Catholic parishioners throughout the United States to understand more clearly the reality of poverty and to support the systemic social changes required to overcome it.

Paralleling the Church's increased attention to social justice throughout the twentieth century was a second movement with profound implications for the life of the Church. That is the liturgical renewal.

9. *Justitia in mundo* (JM), 6.
10. See JM, 18.
11. JM, 34.
12. *Gaudium et spes* (GS), 90.

In this period there were major voices both in Europe and in the United States articulating the nature and direction of this renewal. At a 1909 conference called by Pope Pius X, Belgium theologian Dom Lambert Beauduin called for use of the vernacular language in the peoples' worship. Beauduin, whom many regard as the founder of the twentieth-century liturgical movement in Europe, also argued that the liturgy provides the necessary basis for Christian mission and social outreach.

In the 1930s Fr. Virgil Michel, a monk at St. John's Abbey in Collegeville, Minnesota—and a student of Lambert Beauduin—became a major American voice calling for liturgical reform. Fr. Michel advocated for more active participation of the laity in the liturgy and that the use of the vernacular language was an important step toward that goal.

Virgil Michel believed that lay participation in the liturgy, especially the Eucharist, would lead Catholics to recognize that they are empowered to reform their personal lives as well as the world in which they live. Michel saw a direct connection between liturgy and social justice, between celebrating the Eucharist and working for a more just and caring world. He argued that one cannot enter into the heart of liturgy without being seized with a passion for social justice.

Thirty years later the Second Vatican Council promulgated two documents that advanced both the liturgical renewal and the relationship between liturgy and Catholic social teaching. One of these was the first document issued by this Council; the other was its last. In 1963, the Council fathers approved *Sacrosanctum concilium* (*Constitution on the Sacred Liturgy*). This document opened the doors to a number of changes in liturgical practices that Michel, Beauduin, and many other liturgical reformers had called for—worship in the language of the people, engagement of worshipers in the ritual itself, presider facing the people and inviting them to pray with him.

The *Constitution on the Sacred Liturgy*, like all of the Conciliar documents, drew heavily on biblical and patristic sources, a methodology that favored attention to the personal and historical dimensions of the Christian's life in modern society. The resulting liturgical changes nourished the hope that Catholics taking part in the Eucharist would take their faith into the public square to enact necessary changes in all aspects of life.

The final document that came from the Second Vatican Council was *Gaudium et spes* (*Pastoral Constitution on the Church in the Modern World*). It speaks of a living and mature faith that must prove its "fruitfulness by penetrating the entire life, even the worldly activities, of those who believe, and by urging them to be loving and just, especially towards those in need."[13]

It is tempting to see in *Gaudium et spes* a direct and intentional connection made to *Sacrosanctum concilium*. It is tempting to find in the first and last documents of the Second Vatican Council an explicit linking of liturgy and justice, of liturgy and Catholic social teaching. But such a connection is not made, beyond *Sacrosanctum concilium* simply but importantly recognizing that worship leads us to actions of justice and charity in the world. That more intentional connection would soon appear in postconciliar social teaching.

Liturgy and Catholic Social Teachings: Points of Contact

The connection between liturgy and Catholic social teaching is found in what each calls us to be and to do. Liturgy shows us what our relationship to God often is and what we might hope it to be. In this communal worship, we see our brokenness and our sacred dignity, our need for God's grace and our potential for great acts, our call to ongoing personal conversion and our potential for effecting social change. Through ritual, story, and symbol the liturgy shows us what Catholic social teaching conveys in written documentary form.

Throughout the liturgical year, we hear many readings from the Hebrew prophets admonishing the people—and us—to care for widows and orphans. From other biblical texts we hear the expectation that to be in right relationship with God requires that we care for all of God's creation and that we treat with dignity and love today's immigrants and refugees. The proclamation of the Gospel during Mass reminds all present that Jesus came to free the oppressed, that we are to love our neighbor, and that love is visible in actions such as feeding persons who are hungry and caring for those who are sick. It is hoped that worshippers hearing these biblical messages might recognize that their participation in this liturgy must lead them to certain actions in society—the kind of actions presented through Catholic social teaching.

13. GS, 21.

It is essentially in the Eucharistic liturgy that we see connections to Catholic social teaching. The structure of this ritual reinforces what the social teachings tell us about who we are as human beings. The Introductory Rite has us gathering as one, not as solitary individuals. We worship as a community, a community that at the end of this liturgy is sent forth on mission. The 1971 social document, *Octogesima adveniens*, reflects this point when it notes the Spirit of the Lord "in every place gathering together Christian communities conscious of their responsibilities in society."[14]

The Liturgy of the Word, including the homily, connects this public worship with our life in the world. The Universal Prayer lifts up the needs of this congregation and those of the entire Church, as well as the hurts and challenges facing the various communities to which we belong. This part of the liturgy reminds us, as does the Church's social teaching, that the needs we lift up in prayer become the goals we help to achieve when our liturgy has ended.

Eucharistic Prayer III reflects the Catholic social teaching theme of solidarity when it proclaims: "May this Sacrifice of our reconciliation, / we pray, O Lord, / advance the peace and salvation of all the world." Implicit in this prayer is the recognition that the peace and salvation of the world is not only salvation and freedom from sin but liberation from all other forms of oppression, be they in the economic, social, or political realms. Everyone joining this prayer during the Mass has a role to play in bringing about this salvation and liberation for people anywhere in the world.

If we experience the Eucharist as a thanksgiving for our restored relationships, then we should be able to go forth from this celebration and strive to remove the tensions, strife, and injustices that divide people. By this liturgical act we are empowered to go back into our communities and work to change whatever contradicts what we have just celebrated in the Eucharist.

In their 2002 pastoral reflection, *A Place at the Table*, the United States Catholic bishops touched on this connection between the Eucharist and social justice: "It is Christ's sacrificial meal that nourishes us so that we can go forth to live the Gospels as his disciples. Too often, the call of the Gospel and the social implications of the Eucharist are ignored or neglected in our daily lives."[15]

---

14. OA, 2.
15. *A Place at the Table*, part 1, A.

Three years later, Pope Benedict XVI appeared to summarize this theology in his encyclical, *Deus caritas est* (*God Is Love*). He wrote about the social character of the sacraments and how, when we become one with Christ, we become one with all those to whom Christ gives himself. He then stressed what communion with Christ in the Eucharist must lead to. "A Eucharist which does not pass over into the concrete practice of love is intrinsically fragmented."[16] Further into this document the Holy Father notes that caring for widows and orphans, prisoners, and the sick and needy of every kind is as essential to the Church as the ministry of the sacraments and preaching the Gospel.[17] It is essential to the Church because these acts of charity and justice flow out of the Eucharist.

Many of the Catholic social statements, especially those issued after the Second Vatican Council, develop this relationship between what Christians do at worship and what they do in their public, social lives. For some of these documents that relationship is implicit, perhaps even taken for granted. In others, as already noted, that connection between liturgy and justice is explicit and unmistakable. No other social document is as clear in presenting that relationship between liturgy and Catholic social teaching as *Justitia in mundo*, the 1971 World Synod of Bishops' statement.

In *Justitia in mundo*, the bishops from around the world observe how the liturgy is the "heart of the Church's life"[18] and how it can be a powerful source of education for justice. By this they mean that the liturgy can lead us to understand better our place in the world and motivate us to engage in actions on behalf of justice within that world.

The Eucharist in particular leads us in this direction. It draws us into the act of thanksgiving to God the Father for what has happened to us through the life, death, and Resurrection of Jesus Christ. We offer praise and thanks for having been restored to right relationship with God, with our sisters and brothers, and with the entire creation. *Justitia in mundo* points out that in the Liturgy of the Word and in all the sacraments we discover the teaching of the prophets, of Our Lord, and of the Apostles on the subject of justice.[19] All of the sacraments contribute to the formation of and movement toward justice. The bishops conclude: "Finally, the Eucharist forms the

16. *Deus caritas est* (DCE), 14.
17. See DCE, 22.
18. JM, 58.
19. See JM, 58.

community and places it at the service of all."[20] We go out from this liturgical celebration empowered to engage the task of working to help change whatever contradicts the message of unity, restored relationships, justice, and peace that we have just celebrated.

The Eucharistic meal leads us to engage in social transformation—in acts of justice—because charity and love demand it and because the peace that we proclaim in the Eucharist requires justice. Pope Benedict XVI, reflecting the warning of St. Paul in 1 Corinthians, urges us not to take part in the Eucharist while ignoring the tensions and conflicts in our society and throughout the world.[21] This is the social transformation that the Eucharist both calls us and empowers us to bring about.

As the prophet Micah reminds us, ours is not a God who requires burnt offerings, or animal sacrifices, or rivers of oil. "He has told you, O mortal, what is good; / and what does the LORD require of you / but to do justice, and to love kindness, / and to walk humbly with your God?"[22] This the liturgy proclaims in song, symbol, and ritual. For this task, our Church's social teaching offers moral guidance and practical wisdom for every age.

Purpose of This Book

The liturgy, especially the Eucharist, "is the source and summit of the Christian life."[23] It is where we come together as a people to worship God through rituals that express who we are and how we are to be in relationship with God and with one another. It is a celebration of those relationships once broken and now restored. This liturgy, through use of the ordinary elements of daily life, lets us experience the extraordinary presence of the God we worship, praise, and thank.

Catholic social teaching is intimately related to our celebration of the liturgy on many levels. Most important, these teachings guide us to live as the redeemed and sacred people the liturgy declares us to be. If liturgy is first and foremost an act of worship by the assembled community—by the Church—then this Church's social teachings instruct the worshipers on how they are to live when this public worship is completed.

20. JM, 58.

21. See 1 Corinthians 11:17–33; see also DCE, 14.

22. Micah 6:8.

23. *Lumen gentium*, 11.

As we actively and fully participate in the liturgy, we also gain insights into how we as individuals within this communal context might live out more completely what we celebrate. The liturgy, in other words, shapes our understanding of Catholic social teaching and becomes an important source and influence in the Church's creation of new social documents.

This book is organized around the seven themes of Catholic social teachings. Each of these themes will articulate this connection between liturgy and the social teachings of the Church.

These seven themes were developed in the 1990s by the United States Catholic bishops. Their Task Force on Catholic Social Teaching and Catholic Education asked the question: How well are we doing in familiarizing the Catholic population with our Church's social teachings? Not surprisingly, the answer was "not well."

The Task Force then proceeded to identify resources, programs, and activities that might help Catholic educators design effective approaches to making students at all levels more aware of their Church's social teachings. The Task Force also identified seven key themes that can be found throughout the over 125 years of Catholic social documents. These themes, and the reflection of the bishops following the work of this Task Force, can be found in the 1998 publication, *Sharing Catholic Social Teaching: Challenges and Directions.*

These seven key themes are at the very center of the Catholic social tradition, and they offer a starting point for engaging the Church's social teachings. It will be beneficial to the life of your parish to engage with this resource at liturgy committee, parish council, and peace and justice committee meetings. Use the questions for discussion and reflection at the end of each thematic chapter to evaluate current parish practice to see how well your parish implements the teaching and what needs to be improved (additional questions for young adults are found at https://ltp.org/LCST). Reading papal encyclicals and council documents can be a difficult challenge for most parishioners. Becoming familiar with these seven key themes connects us with the essential messages found in the Catholic social documents. They also help us to live our faith in a way that reflects what we celebrate in our liturgies in a way that guides us in the effort to build a world that more closely reflects the love and mercy, the justice, and the peace of God.

THEME 1

The Life and Dignity of the Human Person

Kevin Ahern

Lord, I am not worthy
that you should enter under my roof,
but only say the word
and my soul shall be healed.

—Invitation to Holy Communion, *The Roman Missal*

AT first glance, the invitation to Holy Communion (noted above) might suggest a sense of dehumanizing self-deprecation. But, this is not the intention. Rather than being humiliating, this short prayer, drawing from the words of the centurion who sought Jesus' help to heal his servant, is one of humility.[1] It helps us to orient ourselves toward God and to acknowledge our place in the order of creation. Like the centurion, Christians know that our reality as created beings is fundamentally different from that of the one true God. But we also know that our reality is one where we are shaped by God's love. Our worthiness to approach the Eucharistic table comes not from our actions but from our very dignity as human beings and the actions of God for us.

The affirmation of the sacredness of the life and dignity of the human person is arguably the core principle of Catholic social teaching. It grounds other principles and specific policy issues, including but not limited to the Church's affirmation of human rights; her teachings on labor justice; her

1. See Matthew 8:8; Luke 7:7.

opposition to elective abortion, euthanasia, and torture; and her teachings on the moral limits of war.

Given its centrality to Catholic social doctrine, how, if at all, does the liturgy relate to the principle of human dignity? This brief chapter explores this dynamic relationship and seeks to offer some ideas for those responsible for the liturgy. But first, what does the Church teach about human dignity?

Human Dignity: The Cornerstone of Catholic Social Teaching

Belief in the inherent dignity of every human person at every stage of life is the cornerstone of the Catholic social tradition. Appeals to this foundational concept can be seen throughout the social encyclicals from *Rerum novarum*, where Pope Leo XIII laments those working "conditions repugnant"[2] to the dignity of workers, to *Laduato Si'*, where Pope Francis shows the connections between a lack of respect for human life and a lack of respect for creation.[3]

In *Gaudium et spes*, the Second Vatican Council outlines two important theological groundings for human dignity. First, drawing from the first creation story in Genesis,[4] the council affirms the sacredness of every person created in the image and likeliness of God, the *imago dei*.[5] Each one of us, by our very nature, has been created good, a social being, who, though wounded by sin, is endowed with intelligence, conscience, and freedom.[6]

Second, the inherent dignity of each human life finds special meaning in the person of Jesus Christ. The Incarnation has profound implications for every one of us. In one of the most beautiful sections of *Gaudium et spes*, the Council writes:

> Human nature, by the very fact that it was assumed, not absorbed, in him, has been raised in us also to a dignity beyond compare. For, by his incarnation, he, the Son of God, has in a certain way united himself with each individual. He worked with human hands, he thought with a human mind. He acted with a human will, and with a human

2. *Rerum novarum*, 36.
3. See *Laudato si'*, 117.
4. See Genesis 1:26–27.
5. See *Gaudium et spes* (GS), 12–17.
6. See GS, 12–17.

> heart he loved. Born of the Virgin Mary, he has truly been made one of us, like to us in all things except sin.[7]

In other words, Christ's Incarnation elevates the dignity of human beings and our activities in the world. Thus, no human being, no part of the human person, and no element in the human experience can ever be understood as completely profane. With the Incarnation, the ineffable God embraces the whole human experience.

To these two theological groundings, we might also affirm a third pneumatological dimension. The human person becomes dignified through the ongoing actions of the Holy Spirit, who transforms us into "temples of God"[8] and who bestows on us ordinary and extraordinary gifts or charisms for the building up of the community. Through these gifts, God is at work in transforming the human experience and our relationship to others.

In sum, the Catholic tradition teaches that every human person possesses dignity regardless of race, class, gender, age, ability, or any other status. This dignity is not distorted by sinful actions or crimes, nor is it lost by age, illness, poverty, or disability. God, as St. John Paul II wrote in *Centesimus annus*, "has imprinted his own image and likeness on"[9] each person, not because of what we do, but because of what we are.

As such, the Church is concerned with any issue that degrades the human person. In the face of political polarizations around life issues, the late archbishop of Chicago, Cardinal Joseph Bernardin, described this perspective as reflecting a "consistent ethic of life," or what the Catholic activist Eileen Egan previously coined as the "seamless garment."[10] The Catholic reverence for life must be holistic and include concerns for abortion, war, racism, injustice, poverty, and capital punishment. This holistic approach to human dignity frames much of Catholic social doctrine. With *Populorum progressio*, Paul VI describes this approach with the concept of integral human development. Here, he affirms a robust understanding of human dignity that looks to the whole person and all people.[11] This holistic vision

7. GS, 22.

8. See 1 Corinthians 3:16.

9. *Centesimus annus*, 11.

10. Joseph Bernardin, "A Consistent Ethic of Life: Continuing the Dialogue, The William Wade Lecture Series, St. Louis University, March 11, 1984," in *The Seamless Garment: Writings on the Consistent Ethic of Life*, ed. Thomas Narn (Maryknoll, NY: Orbis Books, 2008).

11. See *Populorum progressio*, 42.

is reaffirmed by John Paul II and Benedict XVI in *Sollicitudo rei socialis* and *Caritas in veritate.*

A consciousness of human dignity or integral human development, of course, is meaningful only if it shapes our relationships: our relationships with God, with ourselves, and with our neighbors. Referencing the parable of the rich man and Lazarus,[12] *Gaudium et spes* proclaims a particular "reverence" that is owed to each person. All of us, the Council teaches, "should look upon his or her neighbor (without any exception) as another self, bearing in mind especially their neighbor's life and the means needed for a dignified way of life, lest they follow the example of the rich man who ignored Lazarus, who was poor."[13]

In many places in the world, Catholic movements, inspired by the Church's doctrine, play a leading role in bringing about positive social changes. Consider the following examples:

- *Catholic healthcare ministries*, like the hospices run by the Dominican Sisters of Hawthorne, attend to the dignity of the sick and dying.
- *Catholic human rights organizations*, like Pax Romana–ICMICA, advocate for changes in human rights policy in the halls of the United Nations.
- *Catholic prison ministries* draw attention to the dignity of women and men who are too often ignored by society.
- *Catholic humanitarian groups*, like Jesuit Refugee Service and Catholic Charities, offer much-needed social, psychological, and spiritual services to the displaced.

While the rationale may be distinctive, the Church's emphasis on the dignity of the human person is not unique. This value is shared by other cultures and religions. Over the past century, thankfully, the human community has made a number of important steps forward to more fully appreciate human dignity. These include the recognition of women and children as full human beings; people of color and the movements for civil rights; the near abolition of capital punishment, biological warfare, and landmines; the introduction of human rights norms and humanitarian law; the

12. See Luke 16:19–31.
13. GS, 27.

introduction of labor laws; the creation of social security systems; and improvements in administration of medical care. Looking at these developments through the eyes of faith, it is easy to see, as *Gaudium et spes* points out, that God is not absent from these improvements.[14]

Despite some important steps forward, as Pope Francis reminds us in *Evangelii gaudium,* we cannot overlook the massive suffering taking place today. We must, he insists, not forget that, for many, life is precarious: "The joy of living frequently fades, lack of respect for others and violence are on the rise, and inequality is increasingly evident. It is a struggle to live and, often, to live with precious little dignity."[15]

Such situations of *indignity* are supported and sustained by two cultural features frequently identified by Pope Francis. On the one hand, we can see the emergence of an economy of exclusion, in which "human beings are themselves considered consumer goods to be used and then discarded."[16] This has created what Francis describes as a "throwaway culture," a mentality that puts both people and planet in danger.[17] Such exclusion cuts at the heart of the sanctity of life. "Those excluded," he writes, "are no longer society's underside or its fringes or its disenfranchised—they are no longer even a part of it. The excluded are not the 'exploited' but the outcast, the 'leftovers.'"[18]

This economy of exclusion is linked to another cultural feature, what *Evangelii gaudium* describes as the "globalization of indifference." In order to "sustain a lifestyle which excludes others," we become "incapable of feeling compassion at the outcry of the poor, weeping for other people's pain, and feeling a need to help them, as though all this were someone else's responsibility and not our own."[19]

Both exclusion and indifference stand in sharp opposition to Catholic social teaching's emphasis on the dignity of the person and the theological beliefs that ground that doctrine. These social sins of exclusion and indifference manifest themselves in a number of systems of indignity, including, but not limited to, racism and xenophobia, human trafficking, elective

14. See GS, 26.
15. *Evangelii gaudium* (EG), 52.
16. EG, 53.
17. *Laudato si',* 22.
18. EG, 53.
19. EG, 54.

abortion, asymmetrical warfare, and substance abuse. In the face of these challenges, Catholics have a particular responsibility to protect the dignity of each person. Here, the liturgy has an important role to play, but how?

Human Dignity and the Liturgy

As the "source and summit of Christian life,"[20] the liturgical-sacramental life of the Church celebrated throughout the year has much to communicate about human dignity if we pay attention. For example, our communal prayers of praise and thanksgiving speak to the wonders and goodness of God's creation. Moments of confession enable us to see "who we are in God's sight" and to be aware of how our actions and dispositions relate to our own dignity and the dignity of those in our lives.[21] Prayers of intercession enable us to bring the suffering of others, both near and far, into the assembled community. And the proclamation of events in the history of salvation through ritual reenactment and the sharing of the Word remind us of God's abiding love for human beings and our moral responsibility to love God, our neighbors, and ourselves.

Sometimes, however, the linkage between our corporate worship and the dignity of human beings can be hard to perceive, and, if it is visible, it may not be expressed with appropriate depth and breadth. There are two dangers here. On the one hand, the liturgy cannot be reduced only to horizontal social concerns or to one specific social issue. To do so reduces the liturgy to a space of social advocacy, which almost paradoxically robs it of the very power to form socially minded discipleship in the worship of God.

On the other hand, to overlook what the liturgy communicates through Word, song, and symbol concerning the value of human dignity is to miss something central to the very essence of Christian prayer. Archbishop Raymond Hunthausen makes this point clearly: "Liturgy and the reign of God are intimately related. It is impossible to worship God in good conscience and ignore our clear responsibilities to work at building up that

20. *Lumen gentium*, 11.

21. Don E. Saliers, "Liturgy and Ethics: Some New Beginnings," in *Liturgy and the Moral Self: Humanity at Full Stretch; Essays in Honor of Don E. Saliers*, ed. Bruce Morrill and E. Bryon Anderson (Collegeville, MN: Liturgical Press, 1998), 20.

kingdom."[22] As the liturgical theologian Don E. Saliers has pointed out, it was precisely the dissociation of the liturgy from the love of the neighbor that St. Paul critiques in his letter to the community "at Corinth, whose eucharists were a profanization of the body and blood precisely because they did not attend to another."[23]

Racism is a particularly abhorrent example of this failure to attend to the dignity of our sisters and brothers. In her book *Enfleshing Freedom*, theologian M. Shawn Copeland makes this clear by pointing to the contradiction between racism and the Eucharist. "Racism," she writes, "opposes the order of Eucharist. Racism insinuates the reign of sin; it is intrinsic evil. . . . As intrinsic evil, racism is lethal to bodies, to black bodies, to the body of Christ, to Eucharist. Racism spoils the spirit and insults the holy; it is idolatry."[24] Similar contradictions exist between the truth of the Eucharist and the sins of sexism, xenophobia, and classism.

Affirmations of the inherent dignity of each person are expressed in various ways throughout the Church's sacramental life. Here, presiders and liturgical ministers have important opportunities to help make these links more visible over the course of the liturgical year. Five areas are particularly noteworthy.

Human Dignity and the Liturgy of the Word

First, the Church's Lectionary and Liturgy of the Word provide rich opportunities to engage themes of human dignity. Many Scripture readings touch on the sacredness of human life, the Gospel call to love our neighbors, the value of reconciliation, and the goodness of God's creation. When appropriate, homilies, liturgical music, intercessory prayers, and parish programming could help make the connection between revelation and human dignity more explicit.

22. Raymond G. Hunthausen, "Homily for the Opening Session," in *Liturgy and Social Justice: Celebrating Rites, Proclaiming Rights*, ed. Edward M. Grosz (Collegeville, MN: Liturgical Press, 1990), 7.

23. Saliers, "Liturgy and Ethics," 28.

24. M. Shawn Copeland, *Enfleshing Freedom: Body, Race, and Being* (Minneapolis, MN: Fortress, 2010), 109.

Human Dignity and the Liturgy of the Eucharist

Second, the dignity of the human person acquires special significance in the celebration of the Eucharist, a sacrament that speaks in a powerful way to God's love for us, our relationships to others, and our dignity before God. *The Roman Missal* offers several beautiful prayers during the celebration of the Eucharist that uplift human dignity. In the first Eucharistic Prayer for Reconciliation, for example, the celebrant, praises God for being faithful to human beings, despite our shortcomings: "you have bound the human family to yourself / through Jesus your Son, our Redeemer, / with a new bond of love so tight / that it can never be undone."

The fourth Eucharistic Prayer for use in Masses for Various Needs, titled Jesus, Who Went About Doing Good, offers particular insights into the theme of human dignity and the demands of discipleship:

He always showed compassion
for children and for the poor,
for the sick and for sinners,
and he became a neighbor
to the oppressed and the afflicted.

By word and deed he announced to the world
that you are our Father
and that you care for all your sons and daughters. . . .

Open our eyes
to the needs of our brothers and sisters;
inspire in us words and actions
to comfort those who labor and are burdened.
Make us serve them truly
after the example of Christ and at his command.
And may your Church stand as a living witness
to truth and freedom,
to peace and justice,
that all people may be raised up to a new hope.

As the Missal explains, this prayer is appropriately used with a number of the Mass formularies specifically focused on social questions. Some of these social questions include the sanctification of human labor, the plight of refugees and exiles, those experiencing famine, those suffering from hunger, for our oppressors, for those held in captivity or in prison. In the Mass formularies for human labor, for example, the entrance antiphon makes a clear reference to human dignity by quoting Genesis 1, 27, 31: "In the beginning, when God created the heavens and the earth, / God created man in his image; / God looked at everything he had made, and he found it very good." In different ways, these prayers express the Church's integral vision of human development and commitment to the consistent ethic of life.

The most explicit selection of Mass formularies on the theme of human dignity, at least for dioceses in the United States, is For Giving Thanks to God for the Gift of Human Life. While introduced with a focus on ending the practice of elective abortion, the prayers speak to the gift and sacredness of human dignity in its fullness. In one of the two options provided for the Collect, the priest celebrant is invited to pray:

> *O God, who adorn creation with splendor and beauty*
> *and fashion human lives in your image and likeness,*
> *awaken in every heart*
> *reverence for the work of your hands,*
> *and renew among your people*
> *a readiness to nurture and sustain*
> *your precious gift of human life.*

If these Masses for special needs and occasions were used more often, they could go a long way in helping to form the worshiping community on the value of human dignity. Of course, it would be helpful to offer some context as to their significance in homilies or bulletins when these beautiful prayers are used.

Human Dignity and the Sacramental Celebrations

Human dignity is also visible in the celebration of other sacraments, which are very often embodied experiences that engage the sensory, spiritual, and intellectual dimensions of human existence. Through the sacraments of initiation, we are incorporated into the Body of Christ and share in a special way in the profound dignity revealed through the Incarnation. "Rising from the waters of the Baptismal font," as John Paul II wrote, "every Christian hears again the voice that was once heard on the banks of the Jordan River: 'You are my beloved Son; with you I am well pleased.'"[25]

In the Sacraments of the Anointing of the Sick and Reconciliation, we become aware of the holistic and integrated nature of the human person and the goodness of our deepest nature, even when it is wounded by sin or illness. And in the sacraments of Marriage and Holy Orders, we discover that our dignity and relationship with God is enlivened in service to others.

Human Dignity in Specific Celebrations

Specific feast days and holidays offer the liturgical community the chance to incorporate social concerns into our worship of God. Already, the Church in the United States establishes January 22 as "a particular day of prayer for the full restoration of the legal guarantee of the right to life."[26] But there are other connections that can and should be made. Specific feast days of saints could open up concerns about issues of human dignity. The optional Memorial of St. John XXIII (October 11) could address human rights. The optional Memorials of Josephine Bakhita (February 8) or St. Katharine Drexel (March 3) might offer a space to condemn the evils of racism, slavery, and xenophobia. Civic celebrations and some contemporary events could even be connected to liturgical prayer. The Augustinians, for instance, have incorporated the seven international commemorations of the United Nations into their own community calendar. This enables them to recall specific themes, such as racism, the family, and human rights, in their communal worship on those days.

Liturgies Outside the Mass

Last, but not least, the liturgy can help to affirm the sacredness of human life in worship moments outside the Mass. The Mass and sacramental life of the Church are not the only places where Catholic liturgy meets the theme of human dignity. For example, Catholic activists may include moments of prayer in demonstrations against abortion, vigils during an execution, and during demonstrations against racism. The Plowshares

25. *Christifideles laici*, 11.

26. *General Instruction of the Roman Missal*, 373.

Movement, for instance, is known for its creative and controversial use of liturgical elements in their protests against nuclear weapons.[27]

Specific "liturgies for life" can also be celebrated on their own either inside or outside the church building. *Catholic Household Blessings & Prayers* by the United States Bishops' Committee on the Liturgy offers a number of blessings to mark specific moments in life. These include blessings for pregnancy, victims of abuse, and persons with disabilities.

Another rich liturgical repository is the *Book of Blessings*. This text offers a variety of specific blessings for specific needs. Many of these underutilized liturgies speak directly to human dignity in specific moments of life. Among these, two deserve more attention in parish communities today. The loss of a child through miscarriage is a painful experience for parents and family members.[28] The *Book of Blessings* has a beautiful formulation for a blessing after a miscarriage that can be led by a minister or a layperson. The book also includes a specific blessing for those with addictions. In light of the incredible suffering surrounding the present opioid epidemic, specific liturgies might go a long way toward healing.

Before concluding, the topic of dignity in the organization of the liturgy must also be acknowledged. Clearly, there is a fundamental tension if in the celebration and worship of the God of life some members of the community are made to feel less than human. Like the Church as a whole, liturgical ministers, including the non-ordained, must be attentive, as Pope Francis writes in *Evangelii gaudium,* to the dangers of clericalism. This is particularly important, he writes, in relation to women and young people.[29] Unfortunately, there are many stories of people leaving a parish or the Church completely after a dehumanizing or humiliating experience. Being attentive to the dignity of the members of the worshiping community across the liturgical life of the Church is a significant challenge that cannot be overlooked in the Church today.

---

27. See the chapter on Plowshares in Kevin Ahern, *Structures of Grace: Catholic Organizations Serving the Global Common Good* (Maryknoll, NY: Orbis Books, 2015).

28. For more on the power of liturgies following miscarriage, see Susan Bigelow Reynolds, "From the Site of the Empty Tomb: Approaching the Hidden Grief of Prenatal Loss," *New Theology Review* 28, no. 2 (March 2016): 47–59.

29. See EG, 102–5.

Conclusion

In the preface dialogue before the Eucharistic Prayer, the congregation affirms that "it is right and just" to give thanks to the Lord, the source of life and redeemer of the human race. If we are sincere in this proclamation, then we cannot in our prayers and actions be indifferent to any experience that degrades the sacredness of life. Again, this does not mean that we should reduce the liturgy to a litany of issues that threaten human dignity. To expect every liturgy to address all issues of life is too much. At the same time, however, if only a few elements of human dignity get addressed in the worship assembly throughout the year, then something is missing.

The task of proclaiming the integral nature of the human person is not easy, especially in light of the social forces of exclusion and indifference. Thankfully, however, much can be done, both inside and outside the Mass, to form Catholics in what it means to be human and how our communal worship connects to our daily life. To do this requires greater collaboration between liturgical ministers, theological ethicists, and Catholic activists. This project is a welcome step in this direction.[30]

Questions for Parish Staff Discussion and Reflection

- In what ways does our Sunday liturgy affirm the dignity and sanctity of the community as they gather? What does welcoming look and feel like for those who are greeting and for those who are being greeted? Are ministers of hospitality affirmed and trained in their role as those whom worshipers first encounter?
- Does the environment respect the dignity of the elderly and persons with disabilities? Is the lighting, sound, and size of the text in worship aids, hymnals, and bulletins supportive of those with special needs?
- Does the music engage the body, mind, and spirit of the gathered community? Does it affirm each person in their sacred identity as a beloved child of God?
- Does the Universal Prayer regularly incorporate petitions calling for care and respect for the dignity and sanctity of life at every stage and every age?

30. I am grateful to Rev. Thomas Franks, OFM CAP., who served as a dialogue partner.

- On occasion, does the preaching highlight specific issues that reaffirm Catholic social teaching and invite parishioners into active advocacy on behalf of those affected by these issues?
- In what ways does the parish support parishioners who are struggling with a life issue affecting the dignity and sanctity of life—from the unborn to the elderly? Has the parish highlighted common issues affecting many parishioners such as domestic violence, child abuse, elder abuse, workplace injustice and discrimination, and sexual harassment? Has the parish offered resources for helping them? Have you considered forming a ministry in support of victims of domestic abuse?
- Does the parish encourage foster parenting and adoption? Does it support programs that offer individuals and groups a hand up (engendering a sense of dignity) rather than just a handout (helping someone survive another day)?
- Is respect for the dignity and sanctity of life taught in our Catholic schools and religious education programs? Do we incorporate age-appropriate lessons on proper respect and care for body, mind, and spirit and equal regard for the dignity of others? Is theology of the body taught in our Catholic schools and religious education programs and marriage formation?[31]

---

31. Questions for discussion and reflection for young adults may be downloaded for free at https://ltp.org/LCST.

THEME 2

The Call to Family, Community, and Participation

Timothy P. O'Malley

Lord, we implore you:
may these your servants
hold fast to the faith and keep
your commandments;
made one in the flesh,
may they be blameless in all they do;
and with the strength that comes
from the Gospel,
may they bear true witness to Christ
before all.

—Nuptial Blessing, *The Order of Celebrating Matrimony*

IN Catholicism, the family is not reducible to a sociological fact. Instead, as St. John Paul II's Familiaris consortio makes clear, "The Church thus finds in the family, born from the sacrament [of Marriage], the cradle and the setting in which she can enter the human generations, and where these in their turn can enter the Church."[1] The family is a privileged place of

1. *Familiaris consortio* (FC), 15.

communion in which divine love becomes incarnate through the bond of husband and wife, the living out of this bond in the domestic Church, and the family's commitment to Eucharistic solidarity as lived out in the world. It is through the liturgical and sacramental life of the Church that we can see the family for what it is: a living, efficacious sign of divine love in the world.

The Nuptial Bond in the Sacrament of Marriage

Nuptial theology in the West emphasizes Marriage both as a primordial sacrament related to creation and as a conjugal covenant meant to mirror the mutual self-gift of Christ and the Church.[2] That is, Marriage is always both natural and supernatural. The primordial quality to Marriage, grounded in the relationship between man and woman before the fall, indicates the original destiny of man and woman as intended for communion. In contemplating God's original plan for Marriage, one can see the singular harmony that was intended for society. As the Second Vatican Council's *Gaudium et spes* notes: "For God himself is the author of marriage and has endowed it with various values and purposes: all of these have a very important bearing on the continuation of the human race, on the personal development and eternal destiny of every member of the family, on the dignity, stability, peace, and prosperity of the family and of the whole human race."[3] From the faithful concord made possible through the nuptial covenant of man and woman, the human family was to experience the fullness of peace.

Yet, in the first sin of Adam and Eve, there is a wounding of that original nuptial harmony discernable in the first hours of Paradise. When asked by God why he ate from the fruit of the tree of the knowledge of good and evil, Adam responds, "The woman whom you gave to be with me, she gave me the fruit from the tree, and I ate."[4] Adam blames God for the creation of woman, fracturing the unity that man and woman were meant to embody. Once, Adam proclaimed, "This [Eve] at last is bone of my bones / and flesh of my flesh."[5] Now, Adam introduces violence into the world, a disharmony

2. See *The Order of Celebrating Matrimony* (OCM), 1.
3. *Gaudium et spes*, 48.
4. Genesis 3:12.
5. Genesis 2:23.

between man and woman that quickly spills over to their progeny as Cain slaughters Abel.[6] Injustice begins, in the Scriptures, from a fracturing of communion between our first parents: Adam and Eve.

In the New Testament, we discover that Jesus comes to restore this original harmony between man and woman. In the Sermon on the Mount, Jesus demands that his listener fulfill the very heart of the law related to Marriage.[7] Divorce was allowed because of hardness of heart, but in the Kingdom of Heaven, which Jesus comes to inaugurate, there will be no more room for divorce: "Have you not read that the one who made them at the beginning 'made them male and female,' and said, 'For this reason a man shall leave his father and mother and be joined to his wife, and the two shall become one flesh'? So they are no longer two, but one flesh. Therefore what God has joined together, let no one separate."[8] The inauguration of the Kingdom of God begins through a restoration of the original harmony that was to exist between man and woman. Such harmony is not automatic, as Pope Francis has noted, but requires a constant recommitment on the part of the couple: "A love that is weak or infirm, incapable of accepting marriage as a challenge to be taken up and fought for, reborn, renewed and reinvented until death, cannot sustain a great commitment."[9] It is from this natural love, the primordial nature of Marriage as a sacrament, that the entire human family will come to experience the virtues of solidarity, friendship, and sacrifice.

The primordial sacramentality of Marriage is enriched through the Christian economy. As *The Order of Celebrating Matrimony* clarifies:

> Indeed Christ the Lord, making a new creation and making all things new, has willed that Marriage be restored to its primordial form and holiness in such a way that what God has joined together, no one may put asunder, and raised this indissoluble conjugal contract to the dignity of a Sacrament so that it might signify more clearly and represent more easily the model of his own nuptial covenant with the Church.[10]

6. See Genesis 4.
7. See Matthew 5:31–32.
8. Matthew 19:4–6.
9. *Amoris laetitia* (AL), 124.
10. OCM, 5.

In the Sacrament of Marriage, the love of husband and wife, already a reflection of divine communion, is now transformed. The conjugal bond between husband and wife participates "in the mystery of unity and fruitful love between Christ and the Church."[11]

The nature of this bond is made manifest in St. Paul's letter to the Ephesians. Few contemporary couples want to hear about submission of wife to husband in the context of a Marriage. Yet, St. Paul establishes Marriage as a sign of sacrificial love, an act of mutual commitment that consecrates natural love to a supernatural end. Ephesians 5 begins with a liturgical reference, one that exhorts the baptized Christian to become a liturgical offering to God: "Therefore be imitators of God, as beloved children, and live in love, as Christ loved us and gave himself up for us, a fragrant offering and sacrifice to God."[12] To imitate God means to become like Christ, to offer every dimension of human life as a sacrificial offering to the Father. In Marriage, this requires mutual submission of husband and wife to Christ: "Be subject [*hupotassomenoi*] to one another out of reverence for Christ."[13] The Greek verb *hupotasso* does not mean subservience; rather, it is linked in the Scriptures to obedience to divine will. Both husband and wife submit to one another out of reverence for Christ, allowing their act of submission to be taken up into Christ's own loving will. There is an obedience to one another that comes from Christ, a participation in the self-giving love of the Son to the Father. The wife becomes a sign of the Church in the nuptial covenant, submitting to her husband. Yet such submission is not about power. Nor is it reserved solely to the wife. The husband also submits to become the sign of Christ, the Bridegroom, who "loved the church and gave himself up for her."[14] If the fall of man and woman in the Garden was the result of an act of power, of blame, of violence, then Christian marriage is a restoration of this love through the participation of the nuptial bond in the love of the triune God.

In the liturgical act of consent, the ordinary love of husband and wife is infused with the charism of the Holy Spirit. Husband proclaims to his wife:

---

11. OCM, 8.
12. Ephesians 5:1–2.
13. Ephesians 5:21.
14. Ephesians 5:25.

*I, **N.**, take you, **N.**, to be my wife.*
I promise to be faithful to you,
in good times and in bad,
in sickness and in health,
to love you and to honor you
all the days of my life.[15]

The wife proclaims the same to her husband. As Cardinal Marc Ouellet writes about this moment of consent, "From the beginning of their consent, they receive an objective gift of the Spirit (charism), which, touching the intimacy of their conjugal love, transcends their subjectivity and commits them definitively and indissolubly to being credible witnesses of the fidelity of God, who is love."[16] No dimension of the couple's day-to-day living of the mystery of Marriage is to remain unaffected.

The couple becomes a living sign of Christ's sacrificial love. Such love is never private. It is not reduced to sexual desire, to a luxurious wedding that will be featured on TLC, to a vague sense that the other "completes me." The liturgical-sacramental dimension of Marriage reveals that Marriage, just like Christ's sacrifice on the Cross, is given for the renewal of the world. Divine love is a commandment, but in the Sacrament of Marriage, it becomes enfleshed in a relationship between man and woman. As Pope Benedict writes in *Deus caritas est*:

> Love of God and love of neighbor are thus inseparable, they form a single commandment. But both live from the love of God who has loved us first. No longer is it a question, then, of a "commandment" imposed from without and calling for the impossible, but rather of a freely-bestowed experience of love from within, a love which by its very nature must then be shared with others. Love grows through love. Love is "divine" because it comes from God and unites us to God; through this unifying process it makes us a "we" which transcends our divisions and makes us one, until in the end God is "all in all" (1 Corinthians 15:28).[17]

---

15. OCM, 96.

16. Marc Cardinal Ouellet, *Mystery and Sacrament of Love: A Theology of Marriage and the Family for the New Evangelization*, trans. Michelle K. Boras and Adrian J. Walker (Grand Rapids, MI: Eerdmans, 2015), 80.

17. *Deus caritas est* (DCE), 18.

The couple who has received love in the person of their spouse, in the bond of mutual gift that unites them together indissolubly, now offers this love as a gift to the world. Marriage is a participation in an economy of prodigal gift-giving. The love that one has received is to be fruitfully given away as the entire life of the couple has been consecrated anew to God.

The Sacramentality of the Domestic Church

Family life constitutes the primary way that husband and wife give their lives away in love. The life of the family is connected to the possibility of procreation, as the liturgical rites of the Church make clear. Before offering consent, the couple pledges "to accept children lovingly from God / and to bring them up / according to the law of Christ and his Church."[18] In the Nuptial Blessing, the Church intercedes before the Father through the power of Christ, asking:

> *May your abundant blessing, Lord,*
> *come down upon this bride,* ***N.,***
> *and upon* ***N.,*** *her companion for life,*
> *and may the power of your Holy Spirit*
> *set their hearts aflame from on high,*
> *so that, living out together the gift of Matrimony,*
> *they may (adorn their family with children*
> *and) enrich the Church.*[19]

The Church blesses the possibility of procreation, of the gift of children, because human beings cannot manipulate the existence of life. Life always comes as a gift to be received from God. And as Pope Francis notes in *Amoris laetitia*: "The gift of a new child, entrusted by the Lord to a father and a mother, begins with acceptance, continues with lifelong protection, and has as its final goal the joy of eternal life. By serenely contemplating the ultimate fulfillment of each human person, parents will be even more aware of the precious gift entrusted to them."[20] The birth of children to parents is a moment that renews the original nuptial covenant made in Marriage itself,

---

18. OCM, 60.
19. OCM, 209.
20. AL, 166.

restoring husband and wife to the stance of gratitude that should permeate every moment of their existence.

In the Sacrament of Marriage, as Cardinal Ouellet describes, the "spouses' mission is thus to bear witness to an openness to God and to life, which allows the fecundity of the Trinity to be poured out spiritually, physically, and socially in the family."[21] Yet, like Marriage, the Church is aware that the family is not a unique reality within Christian households. Because the family is intrinsic to the original sacramentality of Marriage, one must understand it not merely as a private good but as intrinsic to the well-being of society. As the *Compendium of the Social Doctrine of the Church* clarifies:

> Enlightened by the radiance of the biblical message, the Church considers the family as the first natural society, with underived rights that are proper to it, and places it at the center of social life. Relegating the family "to a subordinate or secondary role, excluding it from its rightful position in society, would be to inflict grave harm on the authentic growth of society as a whole." The family, in fact, is born of the intimate communion of life and love founded on the marriage between one man and one woman. It possesses its own specific and original social dimension, in that it is the principal place of interpersonal relationships, the first and vital cell of society. The family is a divine institution that stands at the foundation of life of the human person as the prototype of every social order.[22]

Catholic social teaching, for this reason, begins not through an abstract articulation of what constitutes the nation-state but through a contemplation of the family. The family reveals the original destiny of human beings as made for communion, for a life composed of social solidarity. It describes how human beings are initiated into this communion through day-to-day life under the same roof, an education of the whole person that takes place as much in the kitchen and living room as it does in the classroom. And in the end, it is the family that manifests to the social order as a whole the proper way of understanding society: "A society built on a family scale is the best guarantee against drifting off course into individualism or collectivism, because within the family the person is always at the center of attention as an end and never as a means."[23] Both individualism and collectivism forget

21. Marc Cardinal Ouellet, *Divine Likeness: Toward a Trinitarian Anthropology of the Family*, trans. Philip Miligan and Linda M. Cicone (Grand Rapids, MI: Eerdmans, 2006), 67.

22. *Compendium of the Social Doctrine of the Church* (CSDC), 211.

23. CSDC, 213.

the uniqueness of the person in a way that family life cannot. Catholic social teaching, even when it deals with nonpersonal actors like nation-states and large financial organizations, can never forget the formation into personhood provided by family life. As Pope Francis notes in *Laudato si'*, the crisis in the family can result in the creation of a throwaway culture where parents no longer consider the well-being of future generations in their own decisions.[24]

The guarding of personhood within the family is at the heart of the Church's teaching around procreation, including contraception and the prohibition against artificial reproductive techniques. Normally, these controversial teachings are understood as exclusively part of the Church's sexual doctrine. But this approach forgets that, in the conjugal bond between husband and wife, every aspect of nuptial life, including sexuality, is meant to become part of the liturgical offering to God. Contraception, as Pope John Paul II describes, can lead to an obscuring of the total gift of self (which includes fertility) that is part and parcel of the nuptial mystery: "This leads not only to a positive refusal to be open to life but also to a falsification of the inner truth of conjugal love, which is called upon to give itself in personal totality."[25] This teaching means that human sexuality in the context of marriage cannot be reduced either to procreation or to the experience of unity alone. Instead, sex is open to life because the love that is shared is not about the sating of desire. It is about a fruitfulness that unfolds in the life of a family, in the birth of children who enter into a household where the law of love reigns supreme.

Of course, such fruitfulness is never guaranteed. Children always come as a gift, and there are couples that are unable to procreate for whatever reason. Their family life is not barren. This barrenness is impossible when it is divine love that unites the couple together in the first place. Such couples can be open to adoption and foster care, a necessity in a fallen world where not every family can provide for the material, personal, and spiritual needs of the child.[26] The social doctrine of the Church is still lacking a robust theology of adoption, one that grounds the act of both raising an adopted child and being an adopted child in the conjugal bond itself. The act of consent in the nuptial liturgy is a form of gift giving that reveals openness to

24. See *Laudato si'*, 162.

25. FC, 32.

26. See AL, 180.

life not reducible to biological fertility. At the same time, it makes clear that the Christian family, although based in biology, is not merely a biological reality. As I've written elsewhere:

> The fruitfulness of nuptial love isn't about our success, the replication of our own self-identity through the ages. It is instead a radical uniting of natural love to the supernatural love revealed upon the cross. It is the transformation of every dimension of human life, as lived within the family, into an occasion for self-gift. For every marriage, whether it is blessed with children or not, is a participation in the passion and resurrection of Jesus Christ.[27]

The infertile couple builds the Christian family through living out the sacrifice of love in the world even in the midst of the very real sorrows of being unable to have their own children. The domestic Church comes into existence when the couple pledges their love for one another in Christ, not when they begin to have children.[28]

The Christian family, in particular, assumes a liturgical and sacramental identity in the world. The family is the domestic Church. This claim is not merely a quaint metaphor by which the Church insists on a robust devotional life in the home. Rather, to call the family the domestic Church is to describe how the communion of the Father, the Son, and the Holy Spirit unfold in the most mundane aspects of human life. As Cardinal Ouellet writes, "The Christian family is a witness to the fact that the Glory of Trinitarian communion, which shines on the face of Christ and his Bride, already dwells within the simplest and most concrete realities of life."[29] In the common prayer of the family, as baptized sons and daughters learn to address God as Father, Son, and Holy Spirit, the glory of divine love is made manifest. In the concrete act of reconciliation, lived within family life, the possibility of conversion is revealed. In gathering weekly at the Eucharistic liturgy of the Church, the family is gradually conformed to the sacrificial love of Christ. The family becomes, when it is at its best, a living sign of what the Church is at her deepest level: a communion of total, self-giving love.

---

27. Timothy P. O'Malley, "The Charism of Infertility," https://churchlife.nd.edu/2017/04/04/editorial-musings-the-charism-of-infertility/; accessed October 18, 2017.

28. Ouellet, *Divine Likeness*, 195.

29. Ouellet, *Divine Likeness*, 76.

Eucharistic Solidarity in the Family

The domestic Church, like the entire Church, does not exist for her own sake. In one of the Eucharistic prefaces for the sacrament of Marriage, we hear how the couple restores the entire Church to her deepest Eucharistic identity:

> *In the union of husband and wife*
> *you give a sign of Christ's loving gift of grace,*
> *so that the Sacrament we celebrate*
> *might draw us back more deeply*
> *into the wondrous design of your love.*[30]

In the pledge of love shared by the couple, in the love made incarnate through the birth of children, the Church is reminded that her life is given to her by Christ to be given away in love.

The family, for this reason, finds its deepest identity in the Eucharistic life of the Church. And like the Eucharistic personality of the Church, the family cannot live its own identity without concrete acts of love in the world. For as Pope Benedict XVI warns, "A Eucharist which does not pass over into the concrete practice of love is intrinsically fragmented."[31]

Pope Francis captures this sense of Eucharistic solidarity that abides within the family. He writes:

> Families should not see themselves as a refuge from society, but instead go forth from their homes in a spirit of solidarity with others. In this way, they become a hub for integrating persons into society and a point of contact between the public and the private spheres. Married couples should have a clear awareness of their social obligations.[32]

The Christian family abides within a communion of love not for its own sake but as a gift to the entire social order. Because the family is a space that recognizes the personhood of the infant child, of the disabled brother or sister, of the aging grandparent, it has a particular gift to offer our world: "A married couple who experience the power of love know that this love is called to bind the wounds of the outcast, to foster a culture of encounter

30. OCM, 200.
31. DCE, 14.
32. AL, 181.

and to fight for justice. . . . [O]pen and caring families find a place for the poor and build friendships with those less fortunate than themselves."[33]

The family's living out of justice, of the works of mercy, are a catalyst for evangelization. The family evangelizes not merely through preaching but through cultivating a civilization of love that is at the heart of the Gospel. The home becomes a space of ecclesial hospitality in the world. The family goes out to meet the lonely, the downtrodden, the hungry and thirsty, those who long for love. The family does not perform such acts of mercy out of an abstract ethics they have discerned on their own. Instead, the family offers this gift of love to the world because the family knows that God is love. The family knows that God is love, contemplates divine love in its prayer, experiences divine solidarity in day-to-day life, and wills to offer this love as a return-gift to the triune God.

Conclusion

The Sacrament of Marriage is thus foundational to connecting liturgy and a life of justice. In Marriage, the couple receives the totality of divine love so that in their union they might become this love for the world. The birth or adoption of children by parents, the caring for the downtrodden by every couple, becomes an icon of divine love functioning in the world. The family does not live this liturgical identity in an abstract manner. Instead, the family is a concrete instantiation of Catholic social doctrine as lived out in relationships between husband and wife, brother and sister, grandparent and cousin. And the family, from the very beginning, is a liturgical reality. A liturgical reality that is fed in the Eucharistic life of the Church.

Questions for Parish Staff Discussion and Reflection

- Has the parish considered having whole families be a part of the ministry of hospitality?
- Do you encourage families (in their various forms) to bring the gifts up during the Preparation of the Gifts?
- Does the music affirm the identity of the parish as one family in Christ?

33. AL, 183.

- Does the parish highlight family life at different times during the liturgical year?
- Are there opportunities given for couples to reflect on married life, mothers to reflect on motherhood, and fathers on fatherhood?
- Do homilies occasionally focus on family life and the sacramental nature of Marriage, as well as encourage and challenge parishioners to support couples, single parents, and families in crisis?
- Do homilies focus on the threats to family life and provide a Christian response to these threats? (For example, issues such as domestic violence, verbal and physical violence against children, the crisis in immigration policy that often separates families, and others).
- Does the Universal Prayer incorporate petitions for families in all of their manifestations?
- Does the parish provide particular ministries to support engaged couples, couples in the first five years of Marriage, and Marriage enrichment opportunities?
- Does the parish have an organization supporting Christian families?
- Does the parish provide support for single parents?
- Does the parish provide opportunities for cohabiting couples or couples not married in the Church to gather for support and encouragement toward living in a sacramental Marriage?
- Does the parish provide support for families in the community beyond the parish?
- Is whole family or intergenerational catechesis offered in the parish?[34]

34. Questions for discussion and reflection for young adults may be downloaded for free at https://ltp.org/LCST.

THEME 3

Rights and Responsibilities

Anne Y. Koester

It is truly right and just,
our duty and our salvation,
always and everywhere to give
you thanks,
Lord, holy Father, almighty
and eternal God,
through Christ our Lord.

—Liturgy of the Eucharist, *The Roman Missal*

"WHAT has the liturgy to do with social reconstruction or the social question? Can the liturgy help to give jobs or raise wages? Can there be any connection between the liturgy and the social problem?"[1] Virgil Michel, OSB, social philosopher and pioneer of the modern liturgical movement in the United States, asked these questions in 1935. The world

1. Virgil Michel, "The Liturgy: The Basis of Social Regeneration," *Orate Fratres* 9 (1935): 536. "Social question" in Dom Virgil's day primarily concerned industrialization and related effects, such as individualism and disregard for the common good. Also, at this time, liturgical renewal was linked with what was referred to as "social regeneration" or "social reconstruction" and the goal of restoring a Christian society. As Margaret Mary Kelleher notes, the Second Vatican Council "abandoned such a goal and began to come to terms with the pluralism of the modern world." See "Liturgy and Social Transformation: Exploring the Relationship," *U.S. Catholic Historian* 16, no. 4, *Sources of Social Reform, Part 2* (Fall 1998): 67.

was still reeling from the devastation of World War I; the United States was under the weight of the Great Depression; cutthroat competition in the marketplace undermined the common good; discrimination based on race, gender, ethnicity, and class was widespread; and excessive individualism obscured a social consciousness. It was four years after Pope Pius XI coined the phrase "social justice" in *Quadragesimo anno* (1931), which spurred Michel's questions.

What relationship did Virgil Michel see between the vexing social issues of his day and the Church's liturgy? To begin with, he emphasized that through Baptism we are "intimately united with Christ and through Christ with [one another]."[2] We are initiated into a "common social life, the Christ-life that unites its members into one fold."[3] We become co-responsible, for we are "no longer to ourselves alone but above all to Christ and his cause."[4] Dom Virgil promoted the liturgy as a primary source of formation and development of the Body of Christ.[5] Through active participation in liturgy, we become conscious of our intimate union with Christ and one another. We are tutored in being "other Christs" in the world.

Significantly, Dom Virgil linked participation in the Church's liturgy with the Christian responsibility for matters of social justice:

> [N]o person has really entered into the heart of the liturgical spirit if [they have] not been seized also with a veritable passion for the re-establishment of social justice in all its wide ramifications.[6]

Simply put, for Virgil Michel, social issues were also liturgical issues. As Benedictine Sister Jeremy Hall, who was personally acquainted with Dom Virgil, said, "Father Virgil had the capacity to look at everything around him and gather it all in to the altar."[7]

How might we look through this same lens as we consider human rights and responsibilities? What do we learn through our participation in the liturgy about embodying and furthering these rights and responsibilities?

---

2. Virgil Michel, *The Christian in the World* (Collegeville, MN: Liturgical Press, 1939), 8.
3. Virgil Michel, "Social Aspects of Liturgy," *Catholic Action* 16 (May 1934), 9.
4. *Orate Fratres*, 9 (1935): 243.
5. Virgil Michel, *Our Life in Christ* (Collegeville, MN: Liturgical Press, 1939), 51.
6. "Timely Tracts: Social Justice," *Orate Fratres* 12 (1938): 132.
7. Personal conversation with Jeremy Hall, OSB, 1999.

Evolution of Rights and Responsibilities

In a brief pastoral message in 1991 marking the occasion of the one hundredth anniversary of *Rerum novarum*, the United States Catholic Bishops highlighted six themes of Catholic social teaching, among them, "the rights and responsibilities of the human person":

> *Flowing from our God-given dignity, each person has basic rights and responsibilities.* These include the rights to freedom of conscience and religious liberty, to raise a family, to immigrate, to live free from unfair discrimination, and to have a share of earthly goods sufficient for oneself and one's family. *People have a fundamental right to life and to those things that make life truly human*: food, clothing, housing, health care, education, security, social services, and employment. *Corresponding to these rights are duties and responsibilities*—to one another, to our families, and to the larger society, to respect the rights of others and to work for the common good.[8]

We can locate the foundation for this theme in Pope John XXIII's *Pacem in terris*. When this encyclical was promulgated in April 1963, the Cuban Missile Crisis had ended only months before, the Second Vatican Council was in its first session, and the beloved pope himself was dying of cancer. Pope John XXIII stated at the outset of *Pacem in terris* the fundamental principle that each individual is "truly a person" and, as such, has rights and duties that are "universal and inviolable, and therefore altogether inalienable."[9] Within the document, he specified several rights inherent to every person, including the rights:

- "to bodily integrity and to the means necessary for the proper development of life";[10]
- "to be respected";[11]

8. *A Century of Catholic Social Teaching: A Common Heritage, A Continuing Challenge* (Washington, DC: United States Catholic Conference, 1991), 4; emphasis added. In 1998, the United States bishops added a seventh theme, care for God's creation; see the USCCB document, *Sharing Catholic Social Teaching: Challenges and Directions*.

9. *Pacem in terris* (PT), 9.

10. See PT, 11.

11. See PT, 12.

- "to share in the benefits of culture," including the right to an education;[12]
- "to worship God in accordance with the right dictates of his own conscience, and to profess his religion both in private and in public";[13]
- "to choose for themselves the kind of life which appeals to them," including a family;[14]
- to work, a just wage, and private ownership of property;[15]
- to emigrate and immigrate;[16] and
- to have an "active part in public life" and contribute "to the common welfare" of all citizens.[17]

Pope John XXIII made it clear that these rights are "inextricably bound up with as many duties,"[18] adding that "to claim one's rights and ignore one's duties, or only half fulfill them, is like building a house with one hand and tearing it down with the other."[19] Moreover, having rights involves the duty to implement these rights, since they express the dignity of the person, and the duty to recognize and respect these rights for every person.[20]

Finally, Pope John stressed the responsibility each has to contribute to and promote the common good, in which every citizen has a right to share:[21] "For the common good, since it is intimately bound up with human nature, can never exist fully and completely unless the human person is taken into account at all times."[22]

Church documents since *Pacem in terris* have affirmed and expanded on John XXIII's articulation of rights and responsibilities. In *Gaudium et spes* (1965), the Second Vatican Council noted that God willed for all persons

12. See PT, 13.
13. See PT, 14.
14. See PT, 15.
15. See PT, 18–22.
16. See PT, 25.
17. See PT, 26.
18. PT, 28.
19. PT, 30.
20. See PT, 44.
21. See PT, 56.
22. PT, 55.

to be "one family."[23] It called for an "increasingly greater recognition" of the basic equality of all persons, given that all "are endowed with a rational soul and are created in God's image; [men and women] have the same nature and origin and, being redeemed by Christ, they enjoy the same divine calling and destiny; there is here a basic equality between all and it must be accorded ever greater recognition."[24] The council insisted that the fundamental rights of the person be respected and that "any kind of social or cultural discrimination in basic personal rights on the grounds of sex, race, color, social conditions, language, or religion, must be curbed and eradicated as incompatible with God's design."[25]

Pope John Paul II further developed the theme of rights and responsibilities. In *Sollicitudo rei socialis* (1987), the pope promoted the rights of participation in the social, cultural, economic, and political aspects of life, while also calling for solidarity and "a firm and persevering determination to commit oneself to the common good; that is to say to the good of all and of each individual, because *we are all really responsible for all.*"[26]

In *Christifideles laici* (1988), Pope John Paul II highlighted the inviolability of the person and the right to life as the most basic and fundamental right and the condition for all other rights:

> The inviolability of the person which is a reflection of the absolute inviolability of God, finds its primary and fundamental expression in the *inviolability of human life.* Above all, the common outcry, which is justly made on behalf of human rights—for example, the right to health, to home, to work, to family, to culture—is false and illusory if *the right to life,* the most basic and fundamental right and the condition for all other personal rights, is not defended with maximum determination.[27]

We find in Pope John Paul II's *Evangelium vitae* (1995) further recognition of the "incomparable worth of the human person."[28] He notes the fullness of life to which each person is called because each person shares in the very life of God.[29] And he underscores the "sacred value of human life" and

23. *Gaudium et spes* (GS), 24.
24. GS, 29.
25. GS, 29.
26. *Sollicitudo rei socialist*, 38; emphasis added.
27. *Christifideles laici*, 38.
28. *Evangelium vitae*, 2.
29. See *Evangelium vitae* (EV), 2.

the "right of every human being to have this primary good respected to the highest degree."[30]

In 2004, in a systematic presentation of the foundations of Catholic social doctrine, the Pontifical Council for Peace and Justice reinforced the Church's understanding that "*the roots of human rights are to be found in the dignity that belongs to each human being*, a dignity that is 'inherent in human life and equal in every person.'"[31] Citing John XXIII's *Pacem in terris* and John Paul II's Message for the 1999 World Day of Peace, the Council for Peace and Justice explained the universal, inviolable, and inalienable nature of these rights:

> *Universal* because they are present in all human beings, without exception of time, place or subject. *Inviolable* insofar as "they are inherent in the human person and in human dignity" and because "it would be vain to proclaim rights, if at the same time everything were not done to ensure the duty of respecting them by all people, everywhere, and for all people." *Inalienable* insofar as "no one can legitimately deprive another person, whoever they may be, of these rights, since this would do violence to their nature."[32]

Pope Benedict wrote in *Caritas in veritate* (2009) that too many people are "concerned only with their rights, and they often have great difficulty in taking responsibility for their own and other people's integral development," claiming they "owe nothing to anyone except to themselves."[33] He cautioned that rights "run wild" and demands escalate when rights are detached from duties.[34]

Most recently, in *Laudato si'* (2015), Pope Francis reminds us that "respect for the human person as such, endowed with basic and inalienable rights ordered to his or her integral development," is the principle that underlies the common good and that we as a society are "obliged to defend and promote the common good."[35]

We do not need to look far to know that the need to respect and promote the fundamental rights of the human person identified above is urgent. Consider the forces that threaten, even disregard, the inherent dignity of

30. EV, 2.
31. *Compendium of the Social Doctrine of the Church* (CSDC), 153.
32. CSDC, 153.
33. *Caritas in veritate (CV)*, 43.
34. CV, 43.
35. *Laudato si'*, 157.

the human person and the rights that flow from this. Discrimination, for instance, remains widespread. Even one incident of people being disrespected and demeaned based on their sex, race, ethnicity, social condition, physical or intellectual ability, religion, language, health, or physical appearance is one too many. It restricts the flourishing of the individual and of society as a whole. In many countries, the right to immigrate sparks heated debates and unfair attacks, prompting suspicion about and mistreatment of immigrants, migrants, and refugees. The right to be respected and treated as worthy is obliterated when human beings are trafficked and enslaved. Societies continue to deal with questions about whether people have a "right" to that which supports their fundamental right to life—such as food and clean water, education, clothing, housing, health care, security, social services, and employment. And the rugged individualism that overshadowed a social consciousness in Virgil Michel's day persists. Modern people are quick to claim their individual rights but then too often deny the same rights for others and sidestep their responsibility to the common good.

Inspired by the questions Virgil Michel posed in 1935, we can ask in our time, "What has the Church's *liturgy* to do with all of this?" In what ways does liturgy bring us to deeper insights and call us to action with respect to human rights and responsibilities? What in particular compels us to preserve and advance human rights, while also fulfilling our corresponding responsibilities to one another, our families, our neighbors, and our global society? What in the liturgy can help us overcome the human tendencies to be self-focused and retreat from doing the hard work needed to ensure that the God-given dignity and rights of every person, known and unknown to us, are protected?

The Liturgy as a "Rehearsal Room" for Rights and Responsibilities

We need to begin with a couple of cautions. One, while it might be tempting to analyze the Church's liturgy, perhaps especially our Eucharistic celebrations, to find within the liturgy specific social strategies or political positions on the rights and duties named in the tradition of Catholic social teaching, such an approach would be a distortion of the purpose of liturgy. Rather, liturgy is, as Mark Searle wrote, more of an "enacted parable"; that is, liturgy is intended to "generate insight and to offer a call rather than to impose moral imperatives; or rather, the moral imperative arises from within the person as a free and personal response to the insight that Jesus gives."[36] The liturgical assembly, he says,

> is the place where justice is proclaimed, but it is neither a classroom nor a political rally nor a hearing. It is more like a rehearsal room where actions must be repeated over and over until they are thoroughly assimilated and perfected—until, that is, the acts have totally identified with the part assigned to them.[37]

Two, the meanings we absorb into our bones through the repetition of these actions are inexhaustible. Liturgy is not containable or confined to one set of meanings, for the liturgy itself and the stories and experiences we bring to it are dynamic and mysterious. Further, as much as we might want to try to manipulate or control the liturgy to fit our agendas, the fact is, before the liturgy is the "people's work," it is first God's work. Key for us is to come to liturgy expectant and available, open to the possibility of new insights and the transformation of attitudes and actions through our participation in the rituals.

With this in mind, let us focus on three aspects of the Church's liturgy that rehearse us in the attitudes and dispositions necessary to lift up and defend the rights of the human person and to carry out our reciprocal responsibilities.

God "Loves Us and Delights in Us":[38] The Sacramentality of the Human Person and of the World

As we saw, a person's right to life and the firm belief in the inherently sacred value of this life is threaded throughout the Church's teaching about rights and responsibilities. These life-affirming convictions are further strengthened by relating them to the principle of sacramentality, which is central to the Catholic imagination and the Church's liturgy and sacramental rituals. It is all of one piece. Sacramentality names what we instinctually know—that

36. "Serving the Lord with Justice," in *Liturgy and Social Justice*, ed. Mark Searle (Collegeville, MN: Liturgical Press, 1980), 30–31.

37. "Serving the Lord," 32.

38. Julian of Norwich, as cited in *Showing of Love*, trans. Julia Bolton Holloway (Collegeville, MN: Liturgical Press; London: Darton, Longman and Todd, 2003).

is, we experience the presence and activity of God in and through the things of this world. God's self-communication and self-offering permeates the world and the lives of the people who are created out of God's love and in God's image. To believe in the sacramentality of the world means to believe in the innate goodness and sacredness of the world and all who live in it.

Through our participation in the Church's liturgy, we rehearse seeing the sacramentality of all of humanity and the world in which we live. Our liturgical celebrations stretch our imagination so we can look beyond the social barriers and prejudices that lead to the isolation and denigration of people. As the symbols of our worship together wash over us again and again, new habits of heart are formed. Our vision is sharpened so we recognize the sacred revealed through people and their stories.

So how does this happen? The Church's liturgy can be ever so subtle in shaping us and, at the same time, ever so powerful when we pay attention—when we are *really present*—to the many symbols of our liturgical celebrations. Consider the following familiar symbols of the liturgy that lift up for us the sacramentality of the human person:

- *The assembly*—a gathering of people who are flawed yet graced, who are incomplete but share in the very life of God and reflect God's splendor. In the worshiping assembly, we believe that Christ is truly present. What more needs to be said to affirm the sacred value of human persons—not only those who are part of the baptized assembly, but all of humanity?

- *Actions of hospitality*—the gestures of welcome that communicate the assembly's "big heart" and that *all* are welcome and valued. The liturgy does not discriminate based on our gender, physical or intellectual abilities, the color of our skin, or our economic circumstances. It matters not whether we live in a mansion or under a blanket on the street, work in a high-rise or in the fields, are married or single, of the World War II generation or a baby boomer, a Gen-Xer, a millennial, or younger. By practicing hospitality in our assemblies, we learn to honor one another by making room for the other not only in the assembly but in our lives. By letting others in, we recognize their inherent dignity and express our own.

- *Touch*—an action that we do and that is done to us. We sign and are signed with the Sign of the Cross when we gather for Eucharist, baptize and welcome the unbaptized to the journey of the catechumenate, and are sealed with Chrism. We take our neighbor's hand to offer the Peace of Christ and extend our hands to another to receive the Body and Blood of Christ. We pray with and anoint those who are sick, and we know God's mercy through touch when we seek forgiveness and reconciliation. We join hands with the one we love in Marriage. Through these experiences of touch, we come to know the value and dignity of the human person.

- *Naming*—a seemingly simple act of affirming the presence of the sacred in one another. In our liturgy, we are named; we are given an identity that announces who we are and are called to become—for instance, the Word of the Lord, the Gospel of our Lord Jesus Christ, the Body of Christ, the Blood of Christ. By being named in our worship rituals, we learn that everyone needs to be named, to know they have an identity as one created in God's image.

This is not an exhaustive list. From beginning to end, our liturgical celebrations lead us to see the sacramentality of ourselves and of all people. When we see in this way, the denial of human rights, discriminatory attitudes and practices, the lack of basic necessities to sustain life, the mistreatment of any kind and in any medium (offline and online) of a human being should be intolerable to us.

Baptism and Discipleship

"The Church of God receives you with great joy. / In her name I sign you with the Sign of the Cross of Christ our Savior."[39] These words are spoken along with the gesture of the Sign of the Cross at the beginning of every celebration of infant Baptism. The Church now receives or claims this child on behalf of Christ. In the adult rite of initiation, the claiming comes through the signing of the senses during the Rite of Acceptance into the Order of Catechumens. As the community claims us for Christ—as we are baptized into the Body of Christ—we are also claimed by and for the community. The sacramental process of Baptism inserts us into a network of relationships—intimately united with Christ and with one another in the

39. *Order of Baptism of Children*, 41, 79, 111, 136, 264, and 302 (revised translation, 2020).

Body of Christ. We become, to borrow from Virgil Michel, "co-responsible" for one another. Our identity as people who "live no longer for ourselves but for Christ"[40] is solidified. It is up to us to live this out.

Our liturgical celebrations, which are communal at their core, tutor us in what it means to live as a community. We rehearse what it means to live in relationship with one another. We are challenged to move beyond the "it's all about me" attitude to "it's all about all of us." And this "us" is not limited to those closest to us or to only the Christian community, for the Body of Christ is not turned in on itself. Rather, the "us" is the human community. Our rehearsal of living in relationship with one another in the Body of Christ should broaden our horizons and instill in us a greater social consciousness and commitment to all of humanity to uphold their human rights and fulfill our responsibilities to the common good.

When we gather for liturgy, we need only to look around us to know that the human community that has a claim on us is richly and delightfully diverse. We are different in all sorts of ways, and yet, we have a shared dignity as persons with common needs—for God, human relationships, those essentials that "make life truly human,"[41] to be free from injustices that serve to suppress our becoming fully human. The liturgy serves to deepen our awareness of this, provided we pay attention, of course. For example:

- *We pray in one voice.* We pray in a diversity of languages, dialects, and accents that are joined together in words and song as we praise God in one voice.
- *We process as a people, individual persons but at the same time a community.* We process from our homes, workplaces, schools, or wherever we are to gather as a liturgical assembly. We experience the opening procession walked by a few on behalf of all. We process to reverence the Gospel and bring to the altar our gifts for Eucharist. We process to receive and to become who we are—the Body and Blood of Christ. We process with the grieving when we honor the dead. All is done in unity as a community of faith.
- *When at Mass, we share in a common meal.* Our participation in the Eucharistic meal expresses our oneness in Christ and

40. 2 Corinthians 5:15.
41. *A Century of Catholic Social Teaching*, 4.

> forms and sustains us as Church. We are also rehearsed in realizing our radical equality in the Body of Christ. Our place in the communion procession is not based on status or wealth or anything else; we all have a privileged place in sharing in the very life of Christ and in becoming what we receive—the Body and Blood of Christ.

These are only a few ways in which the liturgy reminds us of the claiming celebrated at our Baptism—claimed for Christ and by and for the community. We do not shed baptismal identity and the claim on us at the church doors, however. We are called to live as people claimed for Christ and by and for not only the Christian community but the human community.

Participation: Our Right and Duty by Reason of Baptism

Look closely at paragraph 14 of the Church's *Constitution on the Sacred Liturgy:*

> The Church earnestly desires that all the faithful be led to that full, conscious, and active participation in liturgical celebrations called for by the very nature of the liturgy. Such participation by the Christian people as "a chosen race, a royal priesthood, a holy nation, God's own people" (1 Pt 2:9; see 2:4–5) *is their right and duty by reason of their baptism.*[42]

This seminal document speaks of both the right and the duty of the baptized to participate in the Church's liturgy in a manner "demanded by the very nature of liturgy." Importantly, the participation called for concerns both what we do in the rituals *and* in the liturgy of the world. Carrying out our rights and duties to participate in the Church's liturgy prepares and compels us to carry out our rights and duties in the liturgy of the world. This will become more apparent as we reflect first on some of our rights and duties in the liturgical events.

In the liturgy, the baptized have a *right* to

- carry out their primary liturgical ministry, which is to be the assembly;
- offer and receive the hospitality of the gathered Church, assured that everyone belongs;

42. *Sacrosanctum concilium*, 14; emphasis added.

- a worship space that is accessible and supports the right to fully, consciously, and actively participate in the rites for all, regardless of physical or intellectual abilities; worship spaces should foster, not in any way restrict, this right;
- take part in the elements of the ritual actions—the songs, acclamations, responses, listening, gestures, postures, processions, touch, taste, and silence;
- experience meaningful and worthy liturgical celebrations that reflect a thoughtful preparation by the assembly, the presider, and other liturgical ministers. The baptized have a right to effective communication of the layers of meaning in liturgy, which is needed to rightly form the Body of Christ. Poorly celebrated liturgy potentially masks the meaningfulness of the experience and inhibits the assembly's engagement with the liturgy's transformative power.

In Baptism, we are also assigned reciprocal duties or responsibilities with respect to participation in the Church's liturgy. Fulfilling these responsibilities is essential to ensuring the right to fully conscious and active participation. We have, for example, the responsibility to

- *show up!* We are best able to carry out our work as the liturgical assembly when we are fully present and eager to learn our assigned task. Liturgy assumes our physical presence and our mindful engagement in the ritual actions. We need to be together. At the same time, let us not fail to recognize the participation of those unable to be physically present due to health issues or other crises. They too show up for liturgy through their ongoing communion with God and interactions with the community—together with the assembly;
- come open to the possibility that through our rehearsal of who we are meant to be, our attitudes and behaviors are transformed;
- join in Christ's self-offering by offering our own lives;
- carry out our mission—our mandate—to be the Body of Christ in the liturgy of the world. At the end of every liturgical celebration, we are dismissed; we are sent. This does not mark our turning

to our "other life" or "other world." *We are sent from the Church's liturgy to the liturgy of the world.* We are, in the words of one of the forms of the dismissal texts, to "go and announce the Gospel of the Lord." The movement from the Church's liturgy to the liturgy of the world is seamless.

This brings us to consider the other dimension of the participation "demanded by the very nature of liturgy." Alongside our right and duty to participate in the Church's ritual events is our right and duty to participate in the *liturgy of the world.* Michael Skelley, writing on Karl Rahner's theology of worship, states that the "primary and original liturgy" is "the human community's ongoing communion and cooperation with God in history."[43] Liturgy is "what happens when we freely immerse ourselves in the abiding, absolute mystery during the great and small moments of life."[44] The Church's liturgy is "one way in which the liturgy of the world is revealed and celebrated," and without it "we would not be able to grasp fully the height and depth, the length and breadth, of the liturgy of the world."[45] Moreover, the Church's liturgy "provides a necessary means of conversion to such deeper participation in the liturgy of the world."[46]

There is no separation between the Church's liturgy and the liturgy of the world. By reason of our Baptism, we have both a right and a responsibility to participate—fully, consciously, and actively—in the rituals of the Church and in the "primary liturgy" of the world.

Questions for Parish Staff Discussion and Reflection

- In the various formal liturgical ministries, do we have mechanisms to call new people into those ministries and to offer a call to those currently in those ministries to consider other forms of ministry?
- When the Church community offers hospitality in food and refreshment, are all church members called in some way to assist?

43. Michael Skelley, *The Liturgy of the World: Karl Rahner's Theology of Worship* (Collegeville, MN: Liturgical Press, 1991), 93.
44. Skelly, The Liturgy of the World, 93–94.
45. Skelly, The Liturgy of the World, 94.
46. Skelly, The Liturgy of the World, 95.

- Are appropriate liturgical songs like "We Are Called," "Share Your Bread with the Hungry," "What You Have Done for the Least of My People," or others chosen on Sundays to highlight and encourage our collective responsibility?
- Are homilies preached that highlight both the privilege and the responsibility of living our faith and sharing the fullness of our gifts? Are connections made to the gifts that are offered at the Mass and our call as disciples to be the Body and Blood of Christ in the world?
- Are homilies preached regarding the community's responsibility to support the various groups in our society and in our world whose rights are being violated, such as immigrants, people of color, women, victims of human trafficking, and others?
- Does the Universal Prayer include petitions that call the community to participate in and respond to the needs of the people in our world?
- Do the members of our parish community understand the meaning of the sharing of the Sign of Christ's peace as an extension of Baptism and the call to discipleship?
- Is there an active peace and justice or faith-in-action ministry in the parish that helps educate and engage parishioners in supporting those in our own community, and beyond our community, whose rights are being violated?
- Are adults preparing for Baptism and adults preparing to have their children baptized catechized about the responsibilities that flow from our Baptism?
- Is the parish community provided formation opportunities to reflect on their Baptism and what it means to renew their Baptism at Easter? Is charity and justice the heart of this formation?[47]

---

47. Questions for discussion and reflection for young adults may be downloaded for free at https://ltp.org/LCST.

THEME 4

The Option for the Poor and Vulnerable

Thomas Massaro, SJ

For you will that our self-denial
should give you thanks,
humble our sinful pride,
contribute to the feeding of the poor,
and so help us imitate you in
your kindness.

—Preface III of Lent, *The Roman Missal*

DEEP and sincere concern for the poor and vulnerable is a perennial feature of the Christian life. From the very beginning of the Church, followers of Jesus modeled their behavior on a Lord and Savior who was himself materially poor and who so evidently favored the outcasts of society. The Gospel accounts testify that Jesus sought out the poor, spent time with those of humble circumstance, and commended their detachment from material things. From the first generation of Christians on, leaders of the community have urged the faithful to fulfill their solemn obligations to those in need. These include the duty to relieve human suffering whenever we can through *charity* (in almsgiving and direct service) and to undertake acts of *justice* (longer-term planning to change sinful structures that cause poverty, including through political advocacy and education). Our love of neighbor must include actions that reflect a commitment to the well-being of all who suffer in our world.

"Preferential" Option for the Poor

While most central themes of Catholic social teaching have been treated at great length in a plethora of official Church documents on social justice over many decades, the "preferential option for the poor and vulnerable" is a noteworthy exception. This precise phrase appeared in no papal encyclical before 1987[1] and in no official Church document at all before 1979.[2] But of course, the absence of this now-familiar phrase does not mean that believers were not being called to make a deep commitment to the needy in society over all those centuries. Indeed, there was never a time when faithful Christians were *not* reminded of their moral obligations to assist the poor. Naturally, some places and times benefitted from especially strong attention to the ethics of wealth and poverty. On occasion, Christians have been challenged by great prophets of charity and justice who urged the faithful to prioritize acts of direct service and social reform. Among the most effective voices have been those who did not hesitate to bring this message of concern for our low-income neighbors into liturgy and worship. Entering a church sanctuary by no means entails leaving behind our duty to assist the poor. Indeed, liturgy can play a key role in shaping and expressing our social concerns—having our hearts and minds shaped for even greater service to those in need. The power of liturgy to move us beyond apathy and to transform our spirit is on full display when matters of wealth and poverty arise.

One outstanding example of a powerful preacher for social justice is St. John Chrysostom, a fourth-century archbishop of Constantinople who is remembered as one of the most challenging prophetic voices of his era. Like Jesus, he was anything but shy about affirming the necessity of sharing wealth with the poor. His repeated denunciations of the distressing gap between the rich and poor of his great city were so unrelenting that he was forcibly thrown out of town by his wealthy parishioners. For Chrysostom, the homily at Mass provided a privileged opportunity to warn against the spiritual dangers of hoarding wealth and to appeal to the affluent to share their superfluous possessions with their less fortunate neighbors. The texts of his sermons that survived witness to a man who loved rich and poor alike and was deeply distressed by deep social divides. Looking back at the distance

1. See *Sollicitudo rei socialis*, 42.

2. See the documents of the Conference of Latin American Bishops held in Puebla, Mexico, in 1979.

of over fifteen hundred years, we might say that St. John Chrysostom was, in his own bold way, urging the adoption of a "preferential option for the poor and vulnerable."

But, in making this claim, we are getting ahead of ourselves. In order to appreciate the full significance of the phrase "preferential option for the poor and vulnerable" and its potential place in liturgy and worship, we need to double back and consider its precise meaning and origin. The key phrase *the poor* is perhaps the best place to start, although even this seemingly straightforward term has the power to spark disagreement and controversy. To start with, there are many ways to define and understand poverty. Some prefer a strictly material and quantifiable standard to identify who is poor (such as measures of annual income or wealth holdings), while others favor a broader measure of poverty (counting as relevant some less tangible factors, such as access to educational opportunities and "social capital" resources that afford upward mobility).

Those who argue further that poverty involves a state of mind or a facet of culture can point for support to the Gospel according to Matthew,[3] where the first Beatitude praises those who are "poor in spirit" rather than simply the materially poor (while Luke has Jesus saying simply "Blessed are you who are poor"[4]). The Christian tradition has subsequently recognized a judicious distinction between *voluntary poverty* (which is praiseworthy when it is embraced as part of a lifestyle of evangelical witness to the value of simplicity) and *involuntary poverty* (a burden that has tragically been the fate of a majority of humans throughout history). The disagreements and distinctions regarding poverty do not end there. It is easy to affirm the desirability of lifting the burden of suffering from so much of humanity, but elusive indeed is agreement on *how* to do so (for example, should we enact redistributive measures to guarantee equality of opportunity, or even of result?) and *who* is responsible to make attendant sacrifices to achieve these ends.

One thing we can easily agree on is that poverty continues to plague our world, with tragic consequences that include starvation and thousands of preventable deaths each day. Even with considerable recent increases in living standards due to broader education (especially for women) and the

3. See especially Matthew 5:3.
4. Luke 6:20.

benefits of technological advances (especially in agriculture and computerization), the path of progress is seriously incomplete. Nearly one in five people on earth today lives on a dollar a day or less; nearly half live on less than two dollars a day. Economic poverty and food insecurity are still realities that haunt a majority of our fellow travelers on earth. The tragedy of material deprivation is compounded by the trend toward increasing *income inequality*—the ever wider chasm between rich and poor in most societies. As wealth becomes more highly concentrated at the top of the economic pyramid, the affluent tend to accumulate political power as well. The enhanced control of the elite over policymaking and the various institutions of society has a pernicious way of restricting opportunities (especially for quality education and access to desirable career paths) for anyone of modest means, no matter how hard they work to meet the needs of their families. Paths to upward mobility are increasingly blocked as society is further polarized between the "haves" and the "have nots." Much recent scholarship has documented the widening of the wealth and income gaps that exacerbate poverty in our world today.

As distressing as this economic report may sound, the full picture of human inequities and deprivations raises further challenges. Even the broadest definition of "the poor" needs to include "the vulnerable," another key component of the phrase under investigation. Having a decent income and even considerable saved assets is no guarantee against a range of threats that regularly victimize millions. The faces of human suffering are many indeed, and the size of one's bank account provides scant measure of one's vulnerability. People are routinely harmed when, for example, they find themselves in zones of armed conflict, or persecuted for their religion or ethnicity, or forced to flee their homelands and join the tens of millions of refugees and displaced persons seeking resettlement or asylum. Members of minority groups still face severe discrimination and inhumane treatment in many societies. Other noneconomic forms of suffering include racial, gender, or lifestyle discrimination. People are also marginalized due to disease and medical conditions they endure (such as AIDS), mental and physical disabilities, addiction, and other sources of social stigma.

Interestingly, in the Gospel accounts Jesus is frequently portrayed as showing special concern and mercy for precisely these categories of the outcast. He feeds the hungry, welcomes the stranger, and heals the sick, possessed, and disfigured—extending a hand that starts the process of reincorporating all of these marginalized persons into the mainstream of society. For this reason, those who speak today of the "preferential option for the poor and vulnerable" easily find in him inspiration and encouragement for their efforts to break down social divisions and extend a hand of assistance to the poor and vulnerable. But one need not be a Christian to share these social concerns and respond to them with generosity and service. Throughout history, people of all faiths or no faith at all have been motivated to engage in heroic efforts to lift up the poor and vulnerable and invite them into new and healthy relationships. There is undeniably a broadly humanistic dimension of the option for the poor, a commendable fellow-feeling that transcends any particular faith perspective and refuses to be limited by denominational lines. Millions of diverse people in all corners of the world have entered into broad-based coalitions to enact an option for the poor and vulnerable, and we are all the better for this felicitous cooperation against social injustice.

While the Roman Catholic community can claim no monopoly on efforts for social justice that will benefit the less privileged, she can take credit for originating the phrase "preferential option for the poor and vulnerable." This term arose in Catholic circles in Latin America in the 1960s, and it is closely associated with the movement called liberation theology. The Peruvian priest Gustavo Gutiérrez, often called "the father of liberation theology," wrote eloquently about the moral obligation to devote oneself to ending the victimhood that accompanies oppressive economic and social systems and to help empower the poor and vulnerable at last to become subject and agents of their own history. At the time Gutiérrez began writing about liberation theology, Latin America was suffering the residual effects—which persist to this day—of a disastrous colonialism that warped its economy, set up a corrupt caste-like social system, and excluded the vast majority from opportunities to escape near servitude. The aspirations of the common people of an entire continent to own the land they worked and to leave behind illiteracy and exploitation were encouraged by ecclesial communities that recognized the Good News of Jesus Christ as a source of liberation. Impressive programs to encourage literacy and local empowerment took hold throughout Latin America in the second half of the twentieth century, including a proliferation of base Christian communities. While the lion's share of the work of advancement will always belong to the newly liberated people themselves, there remains an important role for allies throughout society. These are the people (liberation thinkers simply refer to them as "friends of the poor") who take up the challenge to make "a preferential option for the poor and vulnerable." Two recent popes, John XXIII and Francis, have been so insistent on rededicating the Church to be "a church of and for the poor" that they must be counted among the key ecclesial supporters of the option for the poor.

This brings us to the two remaining components of our phrase: the concept of "making an option" and its qualifier, "preferential." Making an option implies the existence of a free choice. Will an affluent member of society dedicate his or her energies and resources to helping lift up less privileged neighbors? Such devotion requires a break from ordinary ways of proceeding, a challenge to the inertia of the status quo, an interruption of the apathy of everyday life. There really is a decisive choice at stake, and it involves a type of stark conversion of the soul. Will I extend myself and stick out my neck to practice mercy to those in need of assistance and support? Do I have the moral willpower to give deeply of myself, even more than what I consider my surplus?

Liberation theologian Jon Sobrino (a Jesuit priest writing from El Salvador) titled one of his most challenging books *The Principle of Mercy*, and Pope Francis has made the concept of *mercy* a central pivot of his social teachings. Both call attention to the etymological roots of the word *mercy* (*misericordia* in Latin, and similar in Spanish and other Romance languages), which conjures an image of "one's heart going out to the suffering." While it is not uncommon to have a reaction of generosity to someone in need whom we contact directly, it is remarkable when we extend sincere and effective social concern to those we have never met. And, of course, it is not enough to simply come to an internal act of the will or even to mouth rote words of support for the poor and vulnerable. An abstract and merely notional option for the poor is hardly an option at all. Actions are required to make such an option real—whether this means mailing off an envelope with a donation to a worthy cause (a good start), volunteering time to work directly with low-income neighbors (even better), or committing one's particular expertise to organizations that plan long-term solutions to conditions that perpetuate poverty (an excellent comprehensive approach for long-term social improvement). The most complete options are those that combine direct service to, accompaniment with, and advocacy for the poor, as our talents allow, in an overall strategy that prioritizes empowerment of the poor without in any way fostering long-term dependency. When we muster generosity and make sacrifices for those we have never met (not only strangers, but even "distant strangers" with whom we might share little in common), we are indeed practicing an option for the poor and vulnerable that truly reflects a Christian stance of mercy.

When our hearts go out this way in an enactment of the virtue of mercy, we have made a decisive choice to side with the poor, to be part of the solution to the scourge of poverty and marginalization. But what does the word *preferential* add to this portrayal? While the word *preferential* might seem redundant (and indeed is sometimes omitted from the phrase), it underlines a crucial quality of our choices for social change. One potential shortcoming of speaking in terms of "siding with the poor and vulnerable" is the impression we might create that in doing so we are somehow turning our backs on all others. It would certainly be counterproductive to foster such a perception, as the Gospel is innately universal and inclusive in all ways, constituting perennial Good News for all—rich and poor alike. For this reason, we sometimes hear the amended phrase "a preferential but not exclusive option for the poor." Pope John Paul II employed this more precise phrase on many occasions, including in his 1987 encyclical letter *Sollicitudo rei socialis* (*On Social Concern*) and in his 1991 encyclical *Centesimus annus* (*On the Hundredth Anniversary*). A dedication to address inequities and deprivations in society in no way suggests callousness or indifference toward the affluent. Not only is their dignity to be affirmed and their needs served by the Church, but they represent important potential allies in the struggle for social improvement, as the prophetic ministry of St. John Chrysostom attempted to illustrate many centuries ago.

It should be obvious that the preferential option for the poor and vulnerable builds on the elements of Catholic social teaching developed in other chapters of this book—themes such as the dignity of the human person, the call to life in community, solidarity, worker justice, the full range of rights and responsibilities, and creation. Further, this principle is just as relevant in an affluent society (one with a large and dominant middle class, such as the United States) as it is in a poorer and sharply stratified society (featuring especially extreme and longstanding divisions between upper and lower classes, such as Latin America). When the United States Catholic bishops published a pastoral letter on economic justice in 1986, they did not hesitate to invoke the option for the poor as a guiding principle for the ethical conduct of economic life today. The long and instructive text of *Economic Justice for All* (still highly relevant three decades later) mentions the option for the poor about a dozen times and applies this key justice principle squarely to the national economy. In doing so, the bishops embraced its relevance even for a highly developed industrial society with a high standard of living. Among the conclusions the bishops underline are that "*the poor have the single most urgent economic claim on the conscience of the nation*"[5] and "*the fulfillment of the basic needs of the poor is of the highest priority.*"[6] In identifying these priorities and urgent needs, the bishops were issuing a challenge to affluent and middle-class Americans to make an abiding commitment to advancing the well-being of the poor in our midst. Even though American poverty is not as blatantly death-dealing as it is elsewhere in the world, and even though marginalization is not as disastrous here as it is in many other countries, the urgency of adopting a preferential option for the poor and vulnerable is unquestionable even in the context of the world's leading economy.

The United States bishops were not the first or the only Catholic leaders to commit the Church to an option for the poor. As previously mentioned, the Conference of Latin American bishops (which goes by the Spanish acronym CELAM) had earlier taken the bold step of proposing such an ethical principle. The 1968 CELAM meeting in Medellín, Columbia, had introduced this priority (in the very years that liberation theology was bubbling up in that continent), and the final documents of the next CELAM meeting (held in Puebla, Mexico, in 1979) cemented the commitment by including this phrase prominently in its final documents. But all of these regional meetings of Catholic leaders were in essence following the lead set by the worldwide Second Vatican Council, which in 1965 adopted these words in the opening paragraph of its groundbreaking *Gaudium et spes* (*Pastoral Constitution of the Church in the Modern World*):

5. *Economic Justice for All* (EJA), 86.
6. EJA, 90.

> The joys and the hopes, the grief and anguish of the people of our time, especially those who are poor or afflicted, are the joys and hopes, the grief and anguish of the followers of Christ as well.[7]

In identifying the essence of the Church and her mission with the poor and suffering of our world, the delegates at this great Church council were signaling a wholehearted acceptance of the preferential option for the poor and vulnerable even before this phrase had been coined. Pope Paul VI, writing in 1971, urged Christians to take action to assist their less privileged neighbors:

> The Gospel instructs us in the preferential respect due to the poor . . . ; the more fortunate should renounce some of their rights so as to place their goods more generously at the service of others.[8]

The Liturgy and Service to the Poor

Now, over fifty years after the Second Vatican Council sparked these developments, it is perhaps disappointing that the connection between the liturgy of the Church and the preferential option for the poor and vulnerable is so seldom explicitly drawn and discussed. Perhaps as part of a perennial pattern of missed opportunities for dialogue between liturgists and ethicists, the linkages between the Church's public prayer and her principles of social justice have remained largely unexplored. It is not that Catholics do not pray well in assembly, or that our collective commitments to the poor and vulnerable are weak, but only that our deliberate attention to making these connections between prayer and action have somehow languished. Scholars and practitioners can and should do better to bring together these vital elements of Catholic life. The following paragraphs offer some modest suggestions for reviving the place of the option for the poor (and social justice in general) in our liturgical life.

The Church's liturgical texts contain so many obvious invocations of concern for the poor and our mission to advance the sharing of material goods that we often overlook veiled references to a preferential option for the poor. For example, when the Our Father implores, "Give us this day our daily bread," we rarely notice that the petition is in the plural; in a world of

7. *Gaudium et spes*, 1.

8. *Octogesima adveniens*, 23.

chronic malnourishment, our request must be for *all* to receive the daily sustenance they need. There is no place for self-centeredness or callousness in a worthy offering of this prayer. Similarly, every liturgy has a place for prayers of petition—spontaneous or prepared—and prayerful support for the marginalized should be a ubiquitous element of our prayer life. Most liturgies include collections of freewill offerings, and directing a portion of the proceeds to the needy on a regular basis and announcing this purpose raises awareness of our obligation to our struggling neighbors.

Further, Mass begins with an examination of conscience and a confession of sins, but only rarely are we invited to reflect even briefly on our collective guilt for shirking our duty to the disadvantaged or for complicity in unjust social structures that hurt the poor. Mass ends with a dismissal to go out into the world, but all too seldom is the sending forth prefaced with a reminder of our mission to social justice. Lectionary readings from Old Testament prophets (especially Amos and Jeremiah) and Gospel parables redolent with God's special concern for the poor and vulnerable (such as the Good Samaritan in Luke 10[9] and the Last Judgment in Matthew 25[10]) present further invitations to accord prominence to social justice themes in our regular experience of worship.

We could easily multiply these examples of missed liturgical opportunities to call attention to our duty to the poor. Besides these standard *texts* of the Mass, we must also consider the wider *context* in which a given liturgy is celebrated. While current events should not be allowed to hijack the central movement of the sacred liturgy, most people appreciate sensitive efforts to reference weighty world events in the course of a liturgy, as long as the treatment of what is on everybody's mind anyway is neither heavy-handed nor bursting with opinionated content. For example, when Mass is celebrated on the memorial of a saint or blessed who exemplifies service to the poor (such as St. Vincent de Paul, St. Katharine Drexel, countless others), we witness a rich opportunity to call attention to the option for the poor. Whether in the homily or other elements of the Mass, a liturgy on these occasions enjoys the potential to commend the moral virtues and admirable character traits exemplified by these saints who made an option for

---

9. This Gospel is proclaimed on the Fifteenth Sunday in Ordinary Time during Year C.

10. This Gospel is proclaimed on the Solemnity of Our Lord Jesus Christ, King of the Universe during Year A.

charity and justice in their own time and place. Every participant is implicitly invited to do the same, so why not make this challenge more explicit? If those who prepare and preside over the liturgy do not heighten the assembly's thirst for justice and passion for a more equitable world on these privileged occasions, they have forfeited a valuable opportunity to teach about the ways of our compassionate God.

But, of course, it is wise to refrain from placing too heavy a burden of expectation on the homily or other places for instruction in the course of the Mass. A liturgy is certainly not an ethics class. The primary purpose of a homily is not to persuade members of the congregation to take up social activism or to hold up a brief for countercultural ethical principles, however commendable they may be. Rather, it is to make those present better disciples, more firmly committed to following Christ. Where liturgy and social justice teachings can be resources for one another is the very same place where love for God and love for neighbor overlap. Successful common worship helps all present connect these vertical and horizontal dimensions of the Christian life—relating devotion to God and service of our neighbors. A life-serving liturgy is one that transforms the hearts and minds of the congregation as surely as it transforms bread and wine into something more wonderful, something that makes the in-breaking Kingdom of God mysteriously closer to us all.

Skeptics might reply that, as a foretaste of the Messianic Banquet, worship should be, above all, beautiful, and the edifying nature of a liturgy risks compromise when it is turned into an editorial for the latest social cause, however worthy. This, of course, is not a reason to surrender the goal of linking social justice and liturgy—it is merely a reminder to approach this task deliberately and with great care. Many elements of tasteful liturgy can express our solidarity with the poor in entirely appropriate ways that respect the integrity of the prayer experience, including with the selection of visual elements, hymns, and musical settings that energize us for mission to the Church and world.

The supposed split between prayer and social action is a false one. These two essential components of the Christian life can and should go together; indeed, each inspires us to appreciate the other, and each would be fatally impoverished without the other. If we want our liturgy to witness to the fullness of faith and not just a fraction of our life of discipleship, we must rededicate ourselves to make every effort to bring into our liturgical celebrations the social dimension of our faith, especially key social justice principles such as the preferential option for the poor and vulnerable.

Questions for Parish Staff Discussion and Reflection

- Does the music chosen for the Mass, especially the gathering song, highlight inclusion and embrace diversity and call us to serve the poor?
- Does the preacher speak on various issues that affect the poor and vulnerable? Is the preacher willing and prepared to preach on a relevant issue that sits on the hearts and minds of many in the congregation? Is the preacher willing to risk preaching the Gospel on these issues even when he knows that there will be pushback?
- Does the Universal Prayer include petitions that regularly highlight the plight of the poor and the call for the community to respond in outreach and in advocacy?
- Does the presider use the Eucharistic Prayer for Various Needs IV: Jesus, Who Went About Doing Good, which calls the community to a particular concern for the poor and vulnerable?
- Do parish priorities include significant outreach in stewardship of time, talent, and treasure on behalf of the poor? Has the parish done an assessment of this outreach? Is there a balance between outreach and advocacy?
- Does the parish have a social justice or faith-in-action ministry that focuses specifically on education and advocacy on behalf of the poor and vulnerable? Has the parish considered joining a local faith-based community organizing coalition that addresses systemic change on behalf of the poor and vulnerable?
- Are all parishioners aware of opportunities for participation in outreach and advocacy on behalf of the poor and vulnerable?
- Is the Church's teaching regarding the preferential option for the poor taught in Catholic schools and religious education programs? Are children and families provided opportunities to help the poor in the local community and beyond?[11]

11. Questions for discussion and reflection for young adults may be downloaded for free at https://ltp.org/LCST.

THEME 5

The Dignity of Work and the Rights of Workers

Thomas Scirghi, SJ

O God, who through human labor
never cease to perfect and govern
the vast work of creation,
listen to the supplications of your people
and grant that all men and women
may find work that befits their dignity,
joins them more closely to one another
and enables them to serve their neighbor.

—Collect, Mass for the Sanctification of Human Labor, *The Roman Missal*

IN the beginning man worked. Even while Adam enjoyed the pleasure of Paradise, still he had to till the soil.[1] It is only after the Fall that the work of Adam and Eve was tainted with toil and struggle. Nevertheless, in the Judeo-Christian tradition human labor is revered because it provides the means of survival and growth for individuals and for their communities and because it is the means by which we mortals may encounter almighty God.

1. See Genesis 2:15.

The liturgy is all about work. The meaning is in the word itself. "Liturgy" is derived from the Greek word *leitourgia*, which can be translated as "the work of the people." *Leitourgia* is composed of two words: *leiton*, meaning "people" (think of "laity"), and *ergon*, meaning "work" (think of "ergonomic"). In ancient Rome, this *leitourgia* referred to the taxes the Roman people paid to their government. The tax money contributed to the infrastructure of the city. It paid for the building of bridges, roads, and aqueducts. It helped to maintain a standing army as well as other services. This was the work of the Roman people for the empire. Eventually, with the rise of Christianity and the demise of Caesar, the meaning of *leitourgia* was transferred to Christian worship. The prayer and praise by Christians was their work for God, the One who created the world and redeemed their lives. Thus, liturgy came to mean the "work of the people" for God.

In this chapter we want to ask: What is the relationship between work and worship, between human labor and the Christian liturgy? This is to say, how is the labor of men and women in their daily lives related to the praise and glory of God in Sunday worship? We will begin with a brief discussion of the Church's teaching on human work. Here we will focus on the encyclical *Laborem exercens* by Pope John Paul II. Then we will consider how this teaching on human work may be expressed in the celebration of the liturgy. From the ancient principle, according to Prosper of Aquitaine, *legem credendi lex statuat supplicandi*, which is often shortened to *lex orandi lex credendi* (meaning "the rule of prayer determines the rule of faith"), we can ask how the teaching on human work is reflected and expressed in the liturgy. Indeed, if the liturgy is the "source and summit" of the activity of the Church,[2] then how does the sacred liturgy inform our understanding of human work? The discussion will follow three questions: What is the purpose of work? (And why would the Church be interested in a person's job?) What is the relationship between labor and liturgy? How does the liturgy celebrate human labor?

2. See *Sacrosanctum concilium*, 10, and *Lumen gentium*, 11.

"The Gospel of Work": The Church's Teaching on Human Work

We begin with the question: What is the purpose of work? To be sure, the subject of human labor is of great concern to the Church. This is evident with the number of papal documents on the subject, beginning with *Rerum novarum*, published by Pope Leo XIII in 1891. There Leo writes, "The Church should not be so preoccupied with the spiritual concerns of her children as to neglect their temporal and earthly interests."[3] Some moral theologians consider *Rerum novarum* to be the foundational document of modern Catholic social teaching. Following Leo, several popes have published their own encyclicals commemorating the anniversary of his landmark work. Pius XI published *Quadragesimo anno* for the "fortieth anniversary" in 1931, and Paul VI published *Octogesima adveniens* for the "eightieth anniversary" in 1971. John Paul II published *Laborem exercens* in 1981, on the ninetieth anniversary, and then *Centesimus annus* in 1991, for the centenary.

The Church is concerned with the work of men and women because, simply stated, work is an essential component of human dignity. It is through work that humans develop an understanding of their purpose in life and of where they stand in relation to other people and to God. Good work helps to make us fully human and fully alive.

In recent times the Church's teaching on work developed in response to the "Social Question." The question arose in the nineteenth century and concerned the social problems created by the Industrial Revolution. The rise of the machine created an urban proletariat, that is, a class of people who did not possess capital so they needed to sell their labor in order to survive. Many were forced to leave their lands and move to the cities in search of jobs. Pope Leo XIII commented that many of these people saw their lives reduced to slavery. In his words, "to misuse men as though they were things in the pursuit of gain, or to value them solely for their physical powers . . . is truly shameful and inhuman."[4]

The social question asks what the relationship is between the workers and their work. In light of this social problem, reformers protested the plight of the proletariat. Socialists, like Karl Marx and Friedrich Engels, sought to

3. *Rerum novarum* (RN), 28.

4. RN, 20.

overthrow the economic system. Meanwhile, the Church introduced remedies to the existing economic structure, which became known as the "social doctrine of the Church." One such remedy is the guarantee of a "living wage" for all working people. In 1940, the United States bishops proposed that one's wages should be sufficient to support workers and their families. The living wage should meet the present necessities of life as well as anticipate the challenges of unemployment, sickness, old age and death.[5]

The Church finds the value of human labor rooted in Scripture, beginning with the creation of Adam and the command to take dominion over the earth. In the Book of Genesis we read: "The Lord God took the man and put him in the garden of Eden to till it and keep it."[6] Throughout the Bible individuals are described according to their work, for example, they are shepherds, fishermen, midwives, women drawing water, priests, and soldiers. We hear of Paul, a tentmaker; Lydia, who traded in purple goods; Matthew, the tax collector; and Jesus, a carpenter's son. Moreover, Pope Francis comments that many biblical encounters between God and a person occurred while at work.[7] For example, Moses heard God's voice while tending the flocks of his father-in-law,[8] and Jesus called the disciples while they were at work fishing on the Sea of Galilee.[9]

Furthermore, the members of the early Christian community were reminded of the necessity of work to ward off idleness. For example, Paul admonished the Thessalonians:

> For you yourselves know how you ought to imitate us; we were not idle when we were with you, and we did not eat anyone's bread without paying for it; but with toil and labor we worked night and day, so that we might not burden any of you. . . . [W]e gave you this command: Anyone unwilling to work should not eat.[10]

---

5. United States Catholic Bishops, "Statement on Church and Social Order," Feb. 7, 1940, nn. 41–42. See also: RN, 51; *Quadragesimo anno*, 71; *On Human Work*, 89–91, and *Centessimus annus*, 8. Also, for a good discussion on the subject of a living wage, see William P. Quigley, "The Living Wage and Catholic Social Teaching," *America Magazine*, August 28, 2006.

6. Genesis 2:15.

7. See Pope Francis, "Meeting with the World of Work at the Ilva Factory," Genoa, Italy, May 27, 2017.

8. See Exodus 3:1–6.

9. See Matthew 4:18–22.

10. 2 Thessalonians 3:7–10.

Paul goes on to say that those who are idle and refuse to work should be shunned. Throughout the Scriptures, one's occupation is a significant indication of the quality of one's life. It was expected that healthy people would engage in some sort of work in order to support themselves and the greater community.

Work is necessary for physical survival as well as for spiritual fulfillment. We need to work if we are going to survive in the natural world. We earn wages in order to put food on the table and a roof over our heads, to supply ourselves with clothing, health care, and education. These are some of the physical needs for human beings. Moreover, we work not for ourselves alone but to contribute to the greater social economy.

Furthermore, work provides a sense of dignity for people. Pope John Paul II argues that within the production process, provision should be made for workers to be able to know that in their work they are working for themselves, because their work concerns not only the economy but also, and especially, personal values.[11] This is to say that a person's work has a twofold purpose. In the first place, work is objective, as it begins with the person and is directed outward to an external object, such as cultivating the earth and transforming it through agriculture, industry, service, and technology.[12] In this case, a person's work produces something good and useful. Work is also subjective. Recognizing that we are rational beings made in the image and likeness of God, human work fosters self-realization for the workers.[13] The workers recognize themselves as active participants within a community, contributing to its sustenance and growth. Work upholds the dignity of the worker when these two dimensions—the objective and the subjective—are unified. To be clear, work is necessary for maintaining a thriving economy that supports the society. Work is necessary as well for the good of the workers themselves, for through their labor they come to see themselves more clearly as an image of God.

This subjective—or personal—description is what John Paul II calls the "Gospel of Work." For him, Jesus is the prime example of this Gospel because, although he is the Son of God, he devoted much of his life on earth to manual work at the carpenter's bench. The Gospel of Work proclaims that the

11. See LE, 15.

12. See LE, 4 and 5.

13. See LE, 6.

basis for determining the value of human work is not primarily the kind of work being done but the fact that the one who is doing the work is a person. The Gospel of Work includes all types of labor, even what might be considered menial occupations. In the words of John Paul II, "The sources of the dignity of work are to be sought primarily in the subjective dimension, not the objective one. . . . The primary basis of the value of work is man himself who is its subject."[14] Dr. Martin Luther King Jr. expressed this subjective view of work in a poetic manner in his so-called "Street Sweeper Speech":

> If a man is called to be a street sweeper, he should sweep streets even as Michelangelo painted, or Beethoven composed music or Shakespeare wrote poetry. He should sweep streets so well that all hosts of heaven and earth will pause to say, here lived a great street sweeper who did his job well.

The value of human work is determined from the perspective of the workers, whether they are senators, soldiers, or street sweepers.

When these two dimensions—the subjective and the objective—are unified, then the workers truly may "take dominion" over the earth. In the Book of Genesis we read of how God the Creator pronounced that Adam and Eve will have dominion over the planet with all its abundant life.[15] The notion that we mere mortals should take dominion over the earth is problematic if it is interpreted either as conquering the world or as selfishly using the goods of nature for personal gain. The command to exercise dominion should not suggest a colonial conquistador subjecting a native community to his rule and reaping from the land its precious resources to benefit his king. Rather, dominion by human beings must be understood in light of the One who gives the command. In this way we are to see ourselves as "coworkers" with God. The Lord has given humans the charge to care for his domain. In following the two dimensions of work, when we labor in the field of the Lord, we exert dominion over the world, and we take dominion over ourselves as well. Through our work we become coworkers of the Lord, growing in our relationship with God. As this relationship develops, we realize that we are made in God's own image and likeness, and we carry on the work of creation. This relationship is described well in *Gaudium et spes*:

---

14. LE, 6.
15. See Genesis 1:26–31.

> Individual and collective activity, that monumental effort of humanity through the centuries to improve living conditions, in itself presents no problem to believers, it corresponds to the plan of God. Men and women were created in God's image and were commanded to conquer the earth with all it contains and to rule the world in justice and holiness: they were to acknowledge God as maker of all things and refer themselves and the totality of creation to him, so that with all things subject to God, the divine name would be glorified through all the earth.[16]

It is important to consider that the Judeo-Christian tradition has elevated the value of human labor. This stands in stark contrast to the ancient Greeks, for example, for whom manual labor was considered degrading for a free man and was relegated to slaves.[17] Throughout the Old and New Testaments, human beings act as coworkers with God. Again, in the words of *Gaudium et spes,* men and women "can rightly look upon their work as a prolongation of the work of the creator, a service to other men and women, and their personal contribution to the fulfilment in history of the divine plan."[18] Dorothy Day, in her autobiography, *The Long Loneliness*, considers human beings to be "cocreators" with God. She takes this idea from her friend, the philosopher Peter Maurin:

> God is our creator. God makes us in his image and likeness. Therefore we are creators, he gave us a garden to till and cultivate. We become co-creators by our responsible acts, whether in bringing forth children, or producing food, furniture or clothing. The joy of creativeness should be ours.[19]

Note that the word "cocreator" is problematic since it may distort the relationship between God and humanity. Karl Barth once warned that, while human beings may participate in God's activity, this does not mean that they become a kind of co-God. For Barth, the notion of "cocreator" risks trivializing God's creative act. In *Laborem exercens*, John Paul II nuances this term by explaining that "by means of work man shares in the work of creation."[20] Throughout the encyclical we never find the terms "cocreator"

16. GS, 34.
17. R. Preston, "Doctrine of Work," in *The Westminster Dictionary of Christian Ethics*, ed. James F. Childress and John Macquarrie (Philadelphia: Westminster Press, 1986), 666.
18. GS, 34.
19. Dorothy Day, *The Long Loneliness* (New York: Harper, 1952), 227.
20. LE, 25.

or "cocreation," yet John Paul II makes clear that divine creativity and human work are dynamically related.[21]

So, whereas the term "cocreator" presents a problem in how we express the relationship between God and humanity, the term "coworker" finds a place in the language of the liturgy. For example, throughout the *Book of Blessings* we read prayers directed to "coworkers" who serve God in both sacred and secular fields of work. In one such blessing, namely, the Order for the Blessing of Tools or Other Equipment for Work, we find the following petition: "In entrusting to us the dignity of work, you make us your own co-workers in the world. (Let us bless the Lord.)"[22] And in the prayer of blessing from this order of prayer, we hear: "Grant that by devotion to our own work / we may gladly cooperate with you / in the building up of creation."[23]

In summary, work is necessary for human beings for both objective and subjective reasons. Objectively, work supplies the means for the survival of the human race, allowing people to support themselves with basic necessities and to contribute to the greater community. Subjectively, work empowers human beings to recognize their role as coworkers with almighty God in the ongoing creation of the world and to realize their full humanity, which is reflected in the image and likeness of God.

The Practice of the Liturgy: What Is the Relationship between Labor and Liturgy?

The Roman Catholic liturgy provides the means to celebrate our work. We began this chapter by noting that "liturgy" is a form of work, this is to say that Christian worship is the work of the people for God. We engage in our work for the Lord on our day of rest—the Sabbath. To be clear, this "rest" should not suggest a day of passive lounging. Rather, the Sabbath allows for an active time of re-creation, or ongoing creation.

The Lord commanded that people rest from their work: "Observe the sabbath day and keep it holy, as the Lord your God commanded you. Six days you shall labor and do all your work. But the seventh day is a sabbath

21. Samuel Gregg, "Back to the Person: The Anthropology of *Laborem Exercens* and its Significance for the Social Question," Melbourne, Australia: John Paul II Institute for the Study of Marriage and Family, http://www.stthomas.edu/media/catholicstudies/center/ryan/conferences/pdf/Samuel Gregg.pdf; accessed October 1, 2017.

22. *Book of Blessings* (BB), 932.

23. BB, 935.

to the Lord your God; you shall not do any work."[24] The Letter to the Hebrews reminds the early Christians of this command: "And God rested on the seventh day from all his works. . . . A sabbath rest still remains for the people of God; for those who enter God's rest also cease from their labors as God did from his. Let us therefore make every effort to enter that rest."[25] Several centuries later, in the year ad 321, Roman law decreed, in the Edict of Constantine, that, on "the day of the Sun," the judges, the people of the cities, and the various trade corporations would not work. This decree provided the early Christians with time to devote themselves to prayer as well as to service and celebration with their communities.

The Hebrew word *Sabbath* literally means "no work." The importance of the Sabbath rest is described well in the encyclical *Dies Domini*, published by John Paul II in 1998. He explains that the Lord not only commanded the Israelites to maintain the Sabbath but also provided a model for this command. For the Lord himself worked for six days creating the universe, and on the seventh day he rested. For human beings, rest is a companion to work rather than an interruption. Workers need to withdraw from the demanding cycle of their duties. They need time to relax and revive themselves, physically and mentally. They need a day of rest as well so that they may spend time with family and friends in order to maintain and nurture their relationships.

Along with physical and social well-being, the Sabbath provides for the spiritual health of the workers. This day is sacred, explains John Paul II, because it is a way for people to renew an awareness that all of creation is the work of God, lest we forget that God is the Creator on whom everything depends.[26] It is also a day of solidarity, at which time the Christian community gathers around the Eucharist.[27]

According to John Paul II, the Mass provides a regular reminder of our Baptism. As we enter the Church, the touch of holy water recalls the event in which Christian life is born.[28] We are saved, however, not as individuals, alone, but as members of the Body of Christ. It is important that these members come together to express their identity as the *ekklesia*, "the assembly

24. Deuteronomy 5:12–14.
25. Hebrews 4:4, 9–11.
26. See *Dies Domini* (DD), 65.
27. See DD, 3, 5, 7.
28. See DD, 25.

called together by the Risen Lord."[29] The Eucharistic celebration is filled with expressions of our communal gathering. For instance, the Universal Prayer (Prayer of the Faithful) responds to the proclamation of the Scripture: having heard the Word of God and taken it to heart, we express our concern for others, believing that when we respond to the brothers and sisters of the Lord, we respond to him.[30] Also, during the Communion Rite we pray, "*Our* Father," not "*My* Father," mindful of our shared divine parentage.[31] Then we exchange a "kiss of peace," a sign of solidarity with one another. This communion expressed in word and gesture is then made visible and tangible as we approach the table of the Lord. Finally, the Eucharistic celebration does not halt at the church door because Christians are called to bear witness in their daily lives. They return home and to the workplace inspired in their role as coworkers of the Lord.[32]

The Liturgy: How Does the Liturgy Celebrate Human Labor?

Let us now take a close look at the liturgy, and at one part in particular —the Preparation of the Altar and of the Gifts. At this point members of the congregation present to the Lord their "gifts" in the form of bread and wine, along with money. The presentation of the bread and wine displays the ongoing creation of the faithful. In offering these gifts to the Lord, the priest prays: "Blessed are you, Lord God of all creation, / for through your goodness we have received / the bread [wine] we offer you: / fruit of the earth [of the vine] and work of human hands. / It will become for us the bread of life [become our spiritual drink]." The prayer is conflated here, joining the lines for both the bread and wine. The point of the prayer is to give thanks to God the Creator, the One who provides us with the grain and the grape, which we have fashioned into bread and wine, which then becomes our food and drink. We make good use of what God has given us, for the greater glory of God.

In the procession, the gifts of food are followed by the gift of funds. This money, which is taken from the wages earned by our work, is used to meet the needs of the community, in order to maintain the parish as well as to provide

29. DD, 31.
30. See Matthew 25:40; see also DD, 31.
31. Emphasis added.
32. See DD, 45.

for the greater community in the city, the country, and the world. With this collection, we are able to care for immigrants and refugees or for those suffering from a natural disaster, as well as for other social concerns. Our work and wages contribute to the larger community for the common good.

Let us take a pause here to reflect on how the Preparation of the Altar and of the Gifts is conducted on any given Sunday. For instance, sometimes in the middle of the Mass we will see an usher suddenly invite two of the worshipers to "bring up the gifts" to the priest. These two will be handed the vessels—a carafe of wine and a plate of hosts. The usher will point them to the sanctuary and then they will walk forward dutifully. Meanwhile, back in the pews, the people are fumbling through their purses and wallets for the weekly envelope or for a few dollars to drop into a basket. To an outside observer, this scene could give the impression that the "collection" of money is a fee for the upcoming "meal," rather than a contribution to the missionary work of the Church.

The collection should, however, be a poignant expression of the role of the faithful as "coworkers" with Christ. The fruit of our daily labor, being offered in the liturgy, contributes to the ongoing creation of the Lord. To heighten the significance of the collection we could revive the practice of a procession. Now, *procession* is defined as "the act of a line of persons moving along in orderly succession." Consider that two people carrying a plate of hosts and a carafe of wine is hardly a procession. In some parishes, the carrying of the bread and wine is followed by everyone else in the congregation—those who are able to—walking to the sanctuary where they deposit their offerings in one of several large baskets, lining the first step. This procession provides a vivid display of "the work of the people for God," that is, a celebration of the work of the faithful as they cooperate in building up the Kingdom of God. In the procession, we see the union of human work and Christian worship. In the words of Bishop Frank Dewane, "Work is present every day in the Eucharist, whose gifts are the fruit of man's land and labor. A world that no longer knows the value of work does not understand the Eucharist, the true and human prayer of workers."[33]

The Church is mindful of human labor, which is expressed clearly in the liturgy. For example, reflect on the entrance antiphon for the Mass of

33. Most Rev. Frank J. Dewane, Bishop of Venice, California. USCCB "Labor Day Statement 2017," September 4, 2017.

the Sanctification of Human Labor, the Mass celebrated in the United States on Labor Day: "May your favor, O Lord, be upon us and may you give success to the work of our hands." Consider also the many blessings found for various occupations in the *Book of Blessings.* For example, we find blessings for farmers and fishermen, for healthcare workers and librarians, for office and factory workers, and many more. Moreover, almost every profession has a patron saint. For example, Cecilia is the patron of musicians; Peter, the patron of fishermen; Francis of Assisi, the patron of ecologists; Matthew, the patron of bankers; and Veronica, the patron of laundry workers and photographers. On the days of the memorials for these saints, or for another special occasion, a parish or diocese could give special recognition to those employed in their work. A well-known example is the "Red Mass" for those who work in the legal profession, whose patron is Thomas More. This saint is also the patron of "civil workers," a guild that includes sanitation workers. On his feast day perhaps we could include a special blessing for street sweepers.

The Church's liturgy celebrates the inseparable union of human work and faithful worship. Here we acknowledge that men and women are made for work. But, this work is not the labor of slaves. Rather, we work as friends of the Lord,[34] becoming coworkers in the ongoing creation of the universe. In the words of Pope Francis, "[Work] comes from the first command that God gave to Adam. . . . Where there is a worker there is the interest and gaze of the love of the Lord and of the Church."[35] Gathered in the church, the workers bask in the gaze of the Lord's love.

Questions for Parish Staff Discussion and Reflection

- Because parishioners have varying work schedules and may work multiple jobs, it may be difficult for them to attend Mass. Is the parish staff aware of these individuals? Has the staff extended an invitation of outreach to help them find ways to attend Mass?
- Do homilies support the needs of workers and workers' rights and highlight the injustices that are done to workers in the United States and abroad? (Appropriate times are Labor Day, International Workers Day [May 1], and when a Gospel parable focusing on workers is proclaimed).

34. See John 15:15.
35. Pope Francis, op. cit.

- Does the Universal Prayer occasionally include petitions for the unemployed, the underemployed, and those who suffer discrimination in the workplace? Do petitions call on employers and elected officials to support the lives and livelihood of workers and their families?
- Has the parish staff considered inviting a parishioner or someone from a workers' rights organization to offer a reflection on the dignity of work? Some organizations like Interfaith Worker Justice provide speakers on Labor Day weekend. These reflections could take place during a parish mission, evening of reflection, or following the Prayer after Communion at Sunday Masses.
- Does the parish offer support and/or resources for those who are unemployed in the parish and the greater community (résumé writing, interview skills building, job placement connections)?
- Does the parish provide resources regarding workers' rights to parishioners? Does the parish provide resources for parishioners who are struggling with injustice in their workplace?
- Does the parish have a peace and justice ministry or faith-in-action group? Do they help educate the community on the plight of workers both locally and internationally, offering ways to advocate on behalf of them? Do they educate those about economic injustice and ways to advocate on behalf of workers both locally and nationally?[36]

36. Questions for discussion and reflection for young adults may be downloaded for free at https://ltp.org/LCST.

THEME 6

Solidarity

Kate Ward

As we offer you, O Lord,
the sacrifice,
by which the human race
is reconciled to you,
we humbly pray
that your Son himself
may bestow on
all nations
the gifts of unity and peace.

—Prayer over the Offerings, Solemnity of Our Lord Jesus Christ, King of the Universe, *The Roman Missal*

THE Catholic principle of solidarity is often summed up as "making the concerns of the other my own." We pursue solidarity as a means of acting on our deep belief that God made, and continues to make, us one. While we might not often hear the word solidarity at Mass, the reality, the hope, and the duty it describes are reflected throughout the liturgy. Every time we participate at Mass, we are reminded that God made us one and continues to call us to greater solidarity with one another.

Solidarity in the Catholic Tradition

Solidarity means many things in Catholic social teaching. It describes the truth of the unity of the human family; an ethical principle of being guided by that truth; a practice in response to that truth; and a virtue or quality that particular persons or groups might have, through which they live in response to the truth of human unity. Solidarity describes the way we human beings are created by God, as well as the way we are called to live. Solidarity has an "already-but-not-yet" character; whether we know it or not, we are all linked by our common nature as children of God. Yet, we do not always succeed in living out this reality. We are called to live up to our own nature as people in solidarity with one another, people whose deep care for the concerns of others moves us to action.

Pope John Paul II describes solidarity as awakening to our own interdependence with one another. Solidarity helps us see others as persons like ourselves rather than as tools for our use.[1] It is

> a firm and persevering determination to commit oneself to the common good; that is to say to the good of all and of each individual, because we are all really responsible for all. . . . [It is] a commitment to the good of one's neighbor with the readiness, in the gospel sense, to "lose oneself" for the sake of the other instead of exploiting him, and to "serve him" instead of oppressing him for one's own advantage.[2]

Solidarity is not a passing feeling of sympathy or sadness at another's suffering but a consistent stance toward others and their needs.

While one can, in theory, pursue solidarity alone, the Catholic tradition particularly praises communal efforts toward solidarity through which we express our relational human nature. The United States bishops have equated solidarity with "civic commitment," willingness to take action with others in the public sphere on behalf of the common good.[3] John Paul II insists that workers honor the reality that work brings people together when they join together to improve workplace conditions by forming a union.[4] The right to unionize is crucial for the development of solidarity in society and should never be violated.[5] Solidarity should be honored between countries, even between generations. Wealthy nations express solidarity with poorer ones by pursuing trade and other relationships that benefit

1. *Sollicitudo rei socialis* (SRS), 39.
2. SRS, 38.
3. See *Economic Justice for All* (EJA), 66.
4. See *Laborem exercens*, 20.
5. See EJA, 105.

all their citizens.[6] Preserving the environment is an act of solidarity with future generations.[7]

Catholic social teaching consistently insists that solidarity calls us to change both structures and our own hearts. Benedict XVI called state-sponsored forms of support for human needs a form of solidarity. Yet he pointed out that creating institutions does not guarantee solidarity, which requires each person accepting responsibility for one another.[8] That said, it is crucial to remember that solidarity can't be attained through human effort alone but requires God's freely given love.[9]

Solidarity describes a reality that is and is not. Certain painful realities of human life erode the natural solidarity that exists among all human beings. As the United States bishops write, "Sin simultaneously alienates human beings from God and shatters the solidarity of the human community."[10] Extreme inequalities resulting from human choice erode solidarity by creating division among groups of persons in society.[11] On the personal level, attachments to material things can cause us to forget the reality of human solidarity.[12]

Although solidarity acknowledges the fundamental equality and unity of all human persons, it does not demand that everyone be or act the same; quite the contrary. "The unity of the human family does not submerge the identities of individuals, peoples and cultures," wrote Benedict XVI, "but makes them more transparent to each other and links them more closely in their legitimate diversity."[13] Treating everyone the same could have the unintended result of letting inequities and injustices stand. To pursue solidarity, there are times that particular groups of people, those who have been excluded or treated unjustly, need to be given priority and preferential treatment. For example, according to the United States Catholic bishops, affirmative action programs in hiring and education express solidarity with members of marginalized groups.[14] In the same way, while those who are

6. See EJA, 251–57.
7. See *Caritas in veritate* (CV), 49–51.
8. See CV, 25, 11, 38.
9. CV, 19.
10. EA, 33.
11. See EA, 74, 185.
12. See EA, 328.
13. CV, 53.
14. See EJA, 73.

not poor are called to express solidarity with the poor to help them participate fully in the life of society, the poor are called to express solidarity with one another to the same end.[15]

This is how John Paul II explains the different responses solidarity demands from people who are well off and from those who are poor or oppressed:

> Those who are more influential, because they have a greater share of goods and common services, should feel responsible for the weaker and be ready to share with them all they possess. Those who are weaker, for their part, in the same spirit of solidarity, should not adopt a purely passive attitude or one that is destructive of the social fabric, but, while claiming their legitimate rights, should do what they can for the good of all. The intermediate groups, in their turn, should not selfishly insist on their particular interests, but respect the interests of others.[16]

John Paul II approvingly notes that poor people express solidarity when they recognize common cause with one another and advocate to the rest of society for their own good.

For theologian Father Bryan Massingale, solidarity helps both beneficiaries and victims of unequal systems become more fully human through the realization that God made them one. Solidarity demands not only recognizing the suffering of others and working to change it but recognizing the ways others' suffering sometimes enables our own comfort and convenience and standing ready to give that up.[17] John Paul II agrees, writing: "Solidarity demands a readiness to accept the sacrifices necessary for the good of the whole world community."[18]

Although it describes deep linkage and community, solidarity does not call for a false unity where no one is ever challenged. Theologian Kristin Heyer reminds us of the need for conflictual solidarity. In a world where social structures unequally distribute goods like wealth and safety and evils like poverty, discrimination, and premature death, solidarity calls us into conflict with those unjust structures and persons who defend them. The United States bishops promise: "Christian communities that commit

15. EJA, 88, 119.
16. SRS, 39.
17. Bryan N. Massingale, *Racial Justice and the Catholic Church* (Maryknoll, NY: Orbis Books, 2010), 117.
18. SRS, 45.

themselves to solidarity with those suffering and to confrontation with those attitudes and ways of acting which institutionalize injustice, will themselves experience the power and presence of Christ."[19]

In the Catholic tradition, solidarity, as a reality, describes the oneness of the human family as created by God, a beautiful reality that truly exists and yet is not fully realized. Solidarity, as an ethical ideal, calls us to live up to that reality by acting on the realization that the concerns of those not like us are, indeed, our concerns. This can demand different responses from those struggling and those well off and may even invite us to engage in conflict with the unjust structures that treat some persons as less than full members of the human family.

Where Is God Calling Us to Solidarity?

To describe current realities where the practice of solidarity is sorely needed, we need only think about divisions that cause us to see others as unworthy of our concern rather than as fellow members of the human family. When might we be tempted to say "*those* people" instead of "We"? Where do we often commit psychology's "fundamental attribution error," regarding others' hardships as due to the way they are, while justifying the setbacks of those like ourselves as due not to personal flaws but to extenuating circumstances? Two contemporary situations where God loudly calls Christians to greater solidarity are the situation of migrants and refugees and racial injustice.

Migrants and Refugees

Pope Francis frequently invokes solidarity by noting that God asks two questions of the first humans in Genesis: "Where are you?" and "Where is your brother?" He suggests that we don't know the answer to the first question if we cannot give a satisfactory response to the second.[20] He drew on this image in an anguished homily at Lampedusa, an island in Italy that is one of the first stops for refugees risking their lives to flee war and poverty in their native African countries. "Who is responsible for the blood of these

19. EJA, 55.

20. "Selected Quotes of Pope Francis by Subject" (USCCB Department of Justice, Peace and Human Development, March 2017), 232–34, http://www.usccb.org/beliefs-and-teachings/what-we-believe/catholic-social-teaching/upload/pope-francis-quotes1.pdf.

brothers and sisters of ours?" Pope Francis asked, referring to migrants who died attempting to make the crossing. He continued:

> Nobody! That is our answer: It isn't me; I don't have anything to do with it; it must be someone else, but certainly not me. Yet God is asking each of us: "Where is the blood of your brother which cries out to me?" Today no one in our world feels responsible; we have lost a sense of responsibility for our brothers and sisters.[21]

The lost sense of responsibility for our sisters and brothers manifests in many ways. As Pope Francis describes, we might acknowledge that these people are in need but insist that it is someone else's responsibility to help. For example, in the case of migrants and refugees, we might feel that since they may be geographically quite distant from us, their suffering is "not our problem." Worse yet, we sin against solidarity when we seek to limit our sense of our own responsibility for others by convincing ourselves that others do not deserve our concern. For example, if someone entered another country without the official paperwork that would make their migration legal, we might wish to view that person as a criminal and imagine that this absolves us from the solidarity we are called to extend them as our fellow child of God.

Racial Injustice

Borders between nations provide one occasion of sin against solidarity, of making the mistake of thinking that the concerns of others are not our own. Within the United States, racial difference too often provides cover for making that same mistake. Baptist theologian Jennifer Harvey asks white Americans to envision a world in which we truly believe that the concerns of black Americans are our own. In a blog post titled "Dear Parents of White Children (Like Me)," she asks:

> Let's imagine you lived in a world where you woke up everyday knowing your son or daughter was at risk of being killed by people who represent the state; the very people whose sworn duty it was to "protect and serve." . . . And the list of names of the killers who walked free—some even

21. Pope Francis, "Visit to Lampedusa: Homily of Holy Father Francis" (July 8, 2013).

> receiving a hero's welcome—grew longer too. . . . If it was your child, how would you act? What would you do?[22]

Harvey's words are extremely challenging to many white Americans who may want to argue that this or that person deserved their death at the hands of police. The fact that such a response is even possible, that the first and only response is not horror at an untimely, brutal, and public death, illustrates how easy it is to forget the truth of solidarity, the reality that God made us one and deeply responsible for one another. Solidarity demands no less of white Americans than to take Harvey's challenge seriously and imagine our own children, siblings, or spouses as the potential victims of police violence, hyperincarceration, childhood trauma, and other tragedies that disproportionately, although not exclusively, affect African, Latin, and Native Americans. For white United States Catholics, giving an honest answer to Pope Francis' question "Where is your brother?" means responding to the evils that shorten, curtail, and take the lives of our black, Latino/a and Native sisters and brothers with the same horror—and the same resistance—that would drive us if the victims of these evils belonged to our white families.

Remembering that solidarity can take place in conflict—conflictual solidarity—will mean for some of us joining protests on behalf of racial justice. For others, it might mean speaking up when someone says demeaning or dismissive things about members of different racial groups, migrants, or poor people. It means an end to the false belief that somebody can ever deserve a violent, inhuman death, whether at the hands of the state or in the course of a desperate border crossing.

The Trappist monk Thomas Merton wrote, "Our job is to love others without stopping to inquire whether or not they are worthy."[23] When we extend our solidarity only to those we imagine deserve it, we fail to love them as God does.

Solidarity in the Liturgy

There is no higher liturgical expression of solidarity than the celebration of the Eucharist. The Eucharist unites all Catholics and offers the promise of the

---

22. Jennifer Harvey, "Dear Parents of White Children (Like Me), 2," *Formations: Living at the Intersections of Self, Social, Spirit* (blog), June 19, 2017, https://livingformations.com/2017/06/19/dear-parents-of-white-children-like-me-2/.

23. Letter to Dorothy Day, quoted in Stephen Hand, *Catholic Voices in a World on Fire* (2005), 180.

unity of all human beings in love of God. The Eucharistic Prayers note the "already-but-not-yet" nature of human solidarity, evoking the true unity of all humanity even as we pray for greater unity in Christ.

Nearly every Preface to the Eucharistic Prayer expresses the real solidarity of the human family, describing how God's active love bonds humanity together and humanity to God. For example, we thank God because "when your children were scattered afar by sin, / through the Blood of your Son and the power of the Spirit, / you gathered them again to yourself . . . a people, formed as one by the unity of the Trinity."[24] On the Solemnity of the Most Holy Body and Blood of Christ (and votive Masses of the Most Holy Eucharist) we remember how God offers us the Eucharist "so that the human race, / bounded by one world, / may be enlightened by one faith / and united by one bond of charity."[25]

Various prayers stress the need for reconciliation between sinful humanity and God and between groups of humans whom sin has divided. One of the two Eucharistic Prayers for Reconciliation acknowledges, "Though time and again we have broken your covenant, / you have bound the human family to yourself / through Jesus your Son, our Redeemer."[26] *The Roman Missal* provides texts (Eucharistic Prayers and presidential prayers) for use in Masses for Various Needs and Occasions with groups that stand in need of our solidarity, including refugees and exiles, those in prison, and the oppressor.

The liturgy recognizes the role of human action toward greater solidarity, asking God to "give us a kind heart for the needy and for strangers"[27] and to help us develop "an effective love."[28] That said, while the liturgy can and should help motivate us to take action in solidarity, it also never fails to remind us that the fulfillment of solidarity requires God's assistance. For example: "By the working of your power / it comes about, O Lord, / that hatred is overcome by love, / revenge gives way to forgiveness, / and discord is changed to mutual respect."[29] The liturgy consistently reflects that we are

24. Preface VIII of the Sundays in Ordinary Time.

25. Preface II of the Most Holy Eucharist.

26. Eucharistic Prayer for Reconciliation I.

27. Collect, Mass for Various Needs and Occasions: For Refugees and Exiles.

28. Collect, Mass for Various Needs and Occasions: In Time of Famine or For Those Suffering Hunger.

29. Eucharistic Prayer for Reconciliation II.

already one in God's family, and asks for God's help so that we act accordingly: "Grant, in your mercy, / that the members of the human race, / to whom you have given a single origin, / may form in peace a single family / and always be united by a fraternal spirit."[30] In every Mass, we ask that we may become what God has made us.

In the Gospel accounts, Jesus performs concrete actions in solidarity that we can emulate today, from offering food or medical care to those in need, to welcoming social outsiders, including those branded as sinners. Ideally, the Mass provides opportunities to reflect on how we can act in solidarity in our daily lives. The Universal Prayer or Prayer of the Faithful is our opportunity to attend to particular experiences of suffering or oppression. A well-written Prayer of the Faithful captures the tension between expressing our intent to act in solidarity with "those weighed down by various needs"[31] and remembering that we depend on God's grace to empower us to action.

The real solidarity of the human family means that the suffering sister or brother is always already my problem, always already worthy of my concern. The Eucharist is the greatest possible reminder of this, because in it we experience God's grace offered to an unworthy people. "While we still were sinners Christ died for us."[32] How better can we respond to Christ's freely given love than making the concerns of others our own, whether or not we believe they are worthy?

Saints

Whenever we call on the communion of saints, we acknowledge the interconnected solidarity of the human family, past, present, and future. It is difficult to think of a saint who is *not* noteworthy for making the concerns of others her or his own; two holy women of the United States are good examples.

St. Katharine Drexel (1858–1955) felt called to work on behalf of African American and Native American people and founded the religious order of the Sisters of the Blessed Sacrament to carry out that mission. Drexel and her sisters founded educational institutions throughout the United States for Native Americans and African Americans, who were then widely denied

30. Collect, Mass for Various Needs and Occasions: For the Preservation of Peace and Justice.
31. *General Instruction of the Roman Missal*, 69; see also 70.
32. Romans 5:8.

the opportunity to pursue education. As she and her sisters founded schools and universities, including Xavier University in New Orleans, they placed them in the hands of the communities the institutions served. Frequently encountering opposition to their work, even violence, from racist white people, the sisters encountered many opportunities to practice "conflictual solidarity."[33] St. Katharine was deeply inspired by the Eucharist and its promise to make all people one in God. The Mass for her optional memorial on March 3 connects Eucharistic spirituality to solidarity: "you called Saint Katharine Drexel / to teach the message of the Gospel / and to bring the life of the Eucharist / to the Native American and African American peoples; by her prayers and example."[34]

Venerable Henriette Delille (1812–1862), whose cause for sainthood is under consideration, lived in New Orleans during the first half of the nineteenth century. Restrictive racial codes mandated unequal status for enslaved people of color; free people of color, like Delille and her family; and white people. After she was rejected by two orders of Catholic nuns that admitted white women only, Delille formed a religious community for women of color to share prayer life, educate enslaved people, and prepare them to receive the sacraments. Delille's life of serving poor enslaved people, educating them in defiance of local law and Catholic practice, is a powerful and holy example of solidarity.[35]

As John Paul II explained, solidarity for those in power often demands giving up power, while solidarity for those who are oppressed can involve claiming legitimate rights for members of one's own group. Katharine Drexel, a wealthy white woman, demonstrated solidarity by giving her fortune and her life to serve poor black and Native Americans. Henriette Delille demonstrated solidarity with free women of color like herself by leading them into religious community and demonstrated solidarity with enslaved persons, worse off than herself, by offering them education. The lives of these holy women demonstrate the different forms solidarity can take and its

---

33. Cheryl D. Hughes, *Katharine Drexel: The Riches-to-Rags Life Story of an American Catholic Saint* (Grand Rapids, MI: William B. Eerdmans Publishing Company, 2014).

34. Collect, Optional Memorial of Saint Katharine Drexel, Virgin.

35. M. Shawn Copeland, *The Subversive Power of Love: The Vision of Henriette Delille* (New York: Paulist Press, 2008); Cyprian Davis, *Henriette Delille: Servant of Slaves, Witness to the Poor* (Archives of the Archdiocese of New Orleans, 2004).

consistent response of love for those whose situation in life does not reflect the reality that they are precious members of the one human family.

Conclusion

For Catholics, our life's work is to live up to how God intended creation to be—to act on the truth that we are created by God deeply bound to one another across every form of difference. Painful societal divisions, such as indifference to migrants and racial injustice, show us where God is calling us today to deeper solidarity. We can depend on the uniting power of the Eucharist and the example of Jesus and the saints as we work together to achieve this deeper solidarity, to truly become what God has made and continues to make us.

Questions for Parish Staff Discussion and Reflection

- There are many hymns that speak of solidarity in the Lord: "We Are Many Parts," "Table Song," "In Christ There Is No East or West" are a few examples. Does the music ministry incorporate hymns that call people to solidarity with each other as the Body of Christ? Are hymns that highlight care for the common good sung? Does the music ministry prepare, invite, and actively engage the congregation to sing (in solidarity) to give praise and thanks to God? Is the balance of music in the liturgy skewed more toward congregational singing or more toward reflection and introspection? If a closing hymn is sung, does it invite a sense of solidarity that we as a gathered community of disciples go forth as a unified body to act on behalf of the Lord (that is, does it act as a song of sending forth or a song of commissioning)?
- Do homilies encourage both individual holiness and collective holiness as the Body of Christ? Does the preacher commission people from the congregation each week to serve as baptized disciples? Does he remind parishioners that their call to individual holiness is intimately linked to one another and that all should hold each other in prayer during the week and treat each person as Christ?
- Do homilies reflect on and address societal issues that promote and magnify divisions in our communities, in our country, and in our world that threaten unity as a human community?

- Does the preacher take opportunities to highlight the various symbols of the liturgy and reflect on how these symbols unify and bond us together as one faith community in Christ?
- Does the Universal Prayer include petitions that call us to collective prayer and solidarity with those who are excluded and ostracized due to race, ethnicity, gender, sexual orientation, economic class, or disability? Are petitions included for those in power to change their hearts and minds to build bridges of equitable opportunity for all people, especially for the poor and forgotten?
- Does the presider use the Preface and Eucharistic Prayer for Various Needs I, which raises up the call to unity and solidarity?
- Does the parish community have active involvement in the ministries that express solidarity with those who are sick, homebound, or grieving the loss of a loved one? Are there particular groups in the parish who need to know and experience the support of these ministries (for example, elders, young adults, persons with developmental disabilities, single parents, newly married couples, new parents, and so on).
- Are there groups in the local community who need to be aware of and experience the support from their local Catholic parish, such as the mentally ill, undocumented immigrants, those coming home from the military, or those coming home from prison?[36]

36. Questions for discussion and reflection for young adults may be downloaded for free at https://ltp.org/LCST.

THEME 7

Care for God's Creation

Dawn M. Nothwehr, OSF

You are indeed Holy, O Lord,
and all you have created
rightly gives you praise,
for through your Son our Lord
Jesus Christ,
by the power and working
of the Holy Spirit,
you give life to all things and
make them holy,
and you never cease to gather
a people to yourself,
so that from the rising of the sun
to its setting,
a pure sacrifice may be offered
to your name.

—Eucharistic Prayer III, *The Roman Missal*

TODAY the Maker's earth is plagued with global climate change, scarcity of potable water, loss of biodiversity, food insecurity, and unsustainable energy use that negatively affects billions worldwide. The United States, with about 5 percent of the total global population, devours more than 25 percent of the world's resources, and thus, many Americans, insulated by technological, intellectual, financial, and natural wealth, ignore the urgency and dire impact of the ecological devastation on our sisters and brothers across the globe. But our ecological cushion is shrinking. That reality is a wake-up call and the occasion for an ecological conversion to rediscover our rich Catholic eco-theological tradition and engage the hope this treasure trove offers concerning care for God's creation.

Catholic theological, moral, and spiritual resources for this task are vast—as old as the Book of Genesis and as new as Pope Francis' *Laudato si'* (*On Care for Our Common Home*). In this article, space allows only a glimpse of the Church's social teaching tradition and references to sources that expose it more fully. What do we really mean when we speak of "sacramentality" in relation to the "seven sacraments" or the "sacramentality of creation"? Are those related notions? If so, how is that relatedness served by liturgical pedagogy (catechesis or formation)? How is that sacramentality celebrated on September 1, the World Day of Prayer for the Care of Creation, as inaugurated by Pope Francis? If the liturgy "is the source and summit of the Christian life"[1] and what we pray is what we believe and how we should live (*lex orandi, lex credendi, lex vivendi*), then how does the liturgy form our understanding of how we should care for God's creation?

Catholics today identify the words "sacrament," "sacramentality," or "sacramental" with the seven sacraments of the Catholic Church: A sacrament is an outward sign instituted by Christ to give grace, or St. Augustine's phrase, "a visible sign of an invisible grace." All but lost is the patristic or medieval tenet that salvation history began with creation itself and that Christians valued the Church's ritual sacraments as "particular concentrations of the sacramental nature of all creation. Indeed, sacraments were interpreted as the high manifestations of God's presence in the whole cosmos."[2] Over time, Augustine's broader understanding (*sacramentum*) all but vanished from the horizon in Christian ethics and liturgical pedagogy.[3]

Western Catholicism still suffers from a major disconnect within the framework of Christian sacramentality. The result is precious little discussion

1. *Lumen gentium* (LG), 11.

2. Kenan B. Osborne, *Christian Sacraments in a Postmodern World: A Theology for the Third Millennium* (New York: Paulist Press, 1999), 51. Osborne cites E. Kilmartin, "Theology of the Sacraments: Toward a New Understanding of the Chief Rites of the Church of Jesus Christ," Alternative Futures for Worship, vol. 1, *General Introduction*, (Collegeville, MN: Liturgical Press, 1987), 123–75.

3. See St. Augustine, "Letter 55," Saint Augustine: Letters, vol. I, trans. Sister Wilfred Parsons, SND (New York: Fathers of the Church, 1951).

(implicit or explicit) of sacramentality in the documents concerning the environment or ecological issues. Fortunately, renewed efforts continue to recover the broader "principle of sacramentality" and incorporate it into Catholic social teaching and liturgical pedagogy. To understand this principle of sacramentality, let us now review the Church's teaching on creation.

The Development of the Church's Teaching on Creation

Official Teaching

As we recite each Sunday, our God is the "Creator of heaven and earth."[4] Indeed, "God created the world to show forth and communicate his glory."[5] Catholic doctrine asserts that God created the whole world (spiritual and natural), the world is distinct from God and God created the world in freedom to manifest divine goodness and glory.[6]

The *Catechism of the Catholic Church* includes sections on creation, the seventh commandment ("You shall not steal"), and the common good—each of these sections reflect on the treatment and care for God's earth. Concerning creation, the *Catechism* states it is a gift of God but that each element has intrinsic worth that requires human respect.[7] It affirms that "God wills the *interdependence* of creatures"[8] but also emphasizes a more human-centered view.[9] Humans, bearing the *imago Dei*, must care for the earth.[10] The seventh commandment mandates "respect for the integrity of creation."[11] Indeed, plants, animals, and inanimate beings are destined for the "common good."[12] Animals may service human needs, but any abuse of animals diminishes human dignity.[13] The "common good," the *Catechism* states, requires that

4. Nicene Creed.
5. *Catechism of the Catholic Church* (CCC), 319.
6. See Richard P. McBrien, *Catholicism*, rev. and updated ed. (San Francisco, CA: HarperSanFrancisco, 1994), 254–58.
7. See CCC, 300, 302, and 339.
8. CCC, 340.
9. See CCC, 358.
10. See CCC, 373.
11. See CCC, 2415–18.
12. CCC, 2401.
13. See CCC, 2415.

"the good of each individual is necessarily related to the common good."[14] In fact, the common good concerns the life of all.[15]

All things, therefore, are good and have their own rightful autonomy; they are not simply means to a spiritual end.[16] Grace (divine presence) actually enters into and transforms nature (human life to its fullest context). There is no dichotomy between nature and grace. The immanent presence of God in creation, particularly the Holy Spirit, precludes any dualistic understanding of God and nature.

Scripture and the Early Church

Biblical texts concerning the origins of creation, including humans and their role within it, are much more than Genesis 1 and 2.[17] The Bible clearly links the doctrines of creation, the Incarnation, and redemption. This reality signals the moral, social, and spiritual gravitas of those teachings. Some defining texts include Psalm 104; Proverbs 8:27–30; Job 38:1–4; John 1:1–18; and Colossians 1:15–20.

A chief culprit driving the current moral malaise in the relationship between humanity and creation is the erroneous interpretation of Genesis 1:28: "God blessed them and God said to them: Be fertile and multiply; fill the earth and subdue it. Have dominion over the fish of the sea, the birds of the air, and all the living things that crawl on the earth." Proper exegesis demonstrates that this text does not give license for plundering the natural world but rather articulates God's mandate for humans to be creation's guardians. After the great flood, God promised Noah that he would never destroy life on earth. This promise—the Noahic covenant—emphatically reaffirmed the principle of human reverence for *all* of life.[18]

Sacred Scripture reveals a radically relational God who created a world that is also relational. The Scripture texts reveal how humans, God, creatures, the earth, and all cosmic elements must live together in harmony. Creation is God's good gift; humans are unique yet are related to all creatures as guardian caretakers of the cosmos.

---

14. CCC, 1905.

15. See *Gaudium et spes* (GS), 26 and 74.

16. See CCC, 290–92.

17. See my *Ecological Footprints: An Essential Franciscan Guide for Faith and Sustainable Living (Collegeville, MN: Liturgical Press, 2012)*, especially chaps. 1 and 2.

18. See *Ecological Footprints*, 6–7; see also Genesis 9:16.

Early Christians gleaned rich meanings from the Greek word *oikos,* meaning "household or habitat." They understood the earth as "the dwelling place of God," the *oikumē nē* as "the whole inhabited globe." The habitability of the earth links economy, ecology, and ecumenicity. The economy defined the ordering of God's household for its members' sustenance. The *oikonomia tou theo* ("the economy of God") involved the redemptive transformation of the world, brought to us through Christ Incarnate.

St. Augustine held that the book of nature must be read along *with* the Scriptures in order for us to truly know God.[19] St. Francis of Assisi (patron of ecology) in his *Canticle of the Creatures* sang of the entire creation as kin and family; each being in its uniqueness reveals something of the Divine while praising God. And St. Bonaventure emphasized the seriousness of our integral relationship *within* creation for our salvation:

> Open your eyes, alert your spiritual ears, unseal your lips, and apply your heart so that in all creatures you may see, hear, praise, love, serve, glorify and honor God, lest the whole world rise up against you. For the "universe shall wage war against the foolish." On the contrary, it will be a matter of glory for the wise who can say with the prophet: "For you have given me, O Lord, a delight in your deeds, and I will rejoice in the work of your hands. How great are your works, O Lord! You have made all things in wisdom. The earth is filled with your creatures."[20]

Second Vatican Council and Beyond

Many efforts including the preconciliar liturgical movement led to more direct teaching by the Church on the care for God's creation.[21] Both *Lumen gentium*[22] and *Gaudium et spes*[23] specify that essential care must be given to creation. Western Christian teaching on environmental or ecological issues and ethics grew out of discourse on social, political, and economic problems, particularly dire poverty of the "global south." *Lumen gentium* mandates, however: "[Humans] must, then, acknowledge the inner nature and the value of the whole of creation and its orientation to the praise of

19. See St. Augustine, *De Civitatis Dei, book 16.*

20. Bonaventure, *The Journey of the Soul into God,* 1.15 [5:299], in *Bonaventure: Mystical Writings*, ed. Zachary Hayes, OFM (Charlotte, NC: Tan Publishing, 1999) 77.

21. Timothy O'Malley, "Catholic Ecology and the Eucharist: A Practice Approach," *Liturgical Ministry* 20 (Spring 2011): 69.

22. See LG, 36, 41, 48.

23. See GS, 11, 33, 34, 36, 37, 57, 69.

God, as well as its role in the harmonious praise of God."[24] Sadly, this directive has been largely ignored in Catholic social teaching. Talk of the sacramentality of creation was lost to anthropocentrism that stressed creation's instrumental value. Providentially, *Gaudium et spes* limited anthropocentrism, although the document does not explicitly mention the terms "sacrament," "sacramentality," or the "sacramental principle."

In 2004, the rich social doctrine of the Church was collected for the first time into a single systematic work, including teaching on ecological issues, the *Compendium of the Social Doctrine of the Church.* Chapter 10, "Safeguarding the Environment," is the *Compendium's* largest chapter. The chapter has four parts: "Biblical Aspects," "Man and the Universe of Created Things," "The Crisis in the Relationship between Man and the Environment," and "A Common Responsibility." The fourth section stresses the common good and focuses on wholesome use of biotechnologies, environmental and economic relations, and sustainable "new lifestyles."[25] It treats technology positively but mandates careful responsibility. Overall, having a similar developmental path as *Lumen gentium* and *Gaudium et spes*, the *Compendium* holds an anthropocentric outlook on the environment, but it makes no mention of sacramentality of creation.

"Green" Popes

On June 1, 1972, Pope Paul VI presented a three-page address, *A Hospitable Earth for Future Generations*, to the Stockholm Conference on Human Environment. There he set out major themes that his successors would develop, but Paul VI made no mention of the "sacramentality" of creation.

Pope John Paul II's 1987 *Sollicitudo rei socialis* occasioned the earliest discussion of "sacramental" themes in a social encyclical. But there, "sacrament" referred to the *role of the Church* in creating solidarity in a divided world. Of the eight usages of "sacrament" in that document, four refer to the vocational role of the Church as "the sacrament of unity," one refers to the sacrament of Baptism, and three refer to the Eucharist.[26] John Paul II inferred that somehow these "sacraments" effect in the participants a disposition for

24. LG, 36.

25. *Compendium of the Social Doctrine of the Church*, 486–87.

26. See *Sollicitudo rei socialis*, 14, 31, 40, 47 and 48.

a kind of "ecological solidarity." But it is unclear exactly how "sacrament" functions to form this ethical disposition in anyone.

In his message for the 1990 World Day of Peace, Pope John Paul II brought the ecological crisis to the forefront.[27] He addressed the aesthetic value and doxology of nature, linking faith to numerous ecological issues and critiquing the indiscriminate application of science, technology, and economics that contribute to environmental degradation. He declared that consumerism and instant gratification are the root of the ecological crisis. He called for "ecological conversion," which necessitates major lifestyle changes.[28]

In his 1991 encyclical, *Centisimus annus*, John Paul II wrote: "[People lack] that disinterested, unselfish, and aesthetic attitude that is born of wonder in the presence of being and of the beauty which enables one to see *in visible things the message of the invisible God who created them*."[29] Although John Paul II does not attribute St. Augustine, this document presents an Augustinian description of creation as sacramental.[30]

Pope Benedict XVI enacted initiatives for lifestyle changes, sustainability, and moral leadership in 2007. He set about making the Vatican the world's first carbon-neutral state by installing solar panels on the roof of the Vatican buildings and planting trees in Hungary. In 2008, he emphasized that sustainable human development is of "vital importance" for future generations.[31] On March 10, 2008, the Vatican Office of the Apostolic Penitentiary listed "ecological" offences among the "New Forms of Social Sin."[32] Finally, in *Caritas in veritate*, Benedict XVI linked issues of ecology, economics, genuine love, and justice, and he proposed ways to continue true

---

27. John Paul II, "Peace with God the Creator, Peace with All of Creation," World Day of Peace Address, January 1, 1990; available from https://w2.vatican.va/content/john-paul-ii/en/messages/peace/documents/hf_jp-ii_mes_19891208_xxiii-world-day-for-peace.html.

28. John Paul II, General Audience Address, January 17, 2001; available from https://w2.vatican.va/content/john-paul-ii/en/audiences/2001/documents/hf_jp-ii_aud_20010117.html. Also, Post-Synodal Apostolic Exhortation *Pastores Gregis*, October 16, 2003, 70; available from http://w2.vatican.va/content/john-paul-ii/en/apost_exhortations/documents/hf_jp-ii_exh_20031016_pastores-gregis.html.

29. *Centisimus annus*, 37; emphasis added.

30. Following John Paul II, many Catholic bishops' statements emerged, including the United States Conference of Catholic Bishops pastoral letters: *Renewing the Earth* (1991) and *Global Climate Change: A Plea for Dialogue, Prudence and the Common Good* (2001).

31. "Welcoming Celebration by the Young People Address of His Holiness Benedict XVI," Barangaroo, Sydney Harbour, Thursday, July 17, 2008; available from http://w2.vatican.va/content/benedict-xvi/en/speeches/2008/july/documents/hf_ben-xvi_spe_20080717_barangaroo.html.

32. Philip Pullella, "Vatican Lists 'New Sins,' Including Pollution," *Science News*, March 10, 2008; available from http://www.reuters.com/article/us-pope-sins-idUSL10960232008031O.

human development into the twenty-first century. Benedict emphasized that creation is "God's gift to everyone, and in our use of it we have a responsibility towards the poor, towards future generations and towards humanity as a whole."[33]

On June 18, 2015, Pope Francis released his *magnum opus* and summative work, the encyclical *Laudato si'* (*On Care for Our Common Home*). The term "sacrament" is explicitly mentioned in paragraphs 9 and 235. This document is a profound call to conversion to see creation in the light of faith. The goal of the document is "to enter into dialogue with all people about our common home."[34] The goal of the *dialogue*: "I urgently appeal, then, for a new dialogue about how we are shaping the future of our planet. We need a conversation that includes everyone, since the environmental challenge we are undergoing, and its human roots, concern and affect us all."[35] The dialogue is at the heart of the document, but Pope Francis also has a very striking call to conversion for those in the Church as well.

> The ecological crisis is also a summons to profound interior conversion. It must be said that some committed and prayerful Christians, with the excuse of realism and pragmatism, tend to ridicule expressions of concern for the environment. Others are passive; they choose not to change their habits and thus become inconsistent. So what they all need is an "ecological conversion," whereby the effects of their encounter with Jesus Christ become evident in their relationship with the world around them. Living our vocation to be protectors of God's handiwork is essential to a life of virtue; it is not an optional or a secondary aspect of our Christian experience.[36]

United States Bishops

In 1991, the United States bishops contributed to the "wider conversation" of the ecological crisis by issuing their pastoral letter *Renewing the Earth: An Invitation to Reflection and Action on Environment in Light of Catholic Social Teaching.* In this letter, the bishops explicitly describe the universe as "sacramental."[37] The bishops state: "*[A] God-centered and sacramental view of the*

33. *Caritas in veritate* (CV), 48.
34. *Laudato si'* (LS), 3.
35. LS, 14.
36. LS, 217; emphasis added.
37. See *Renewing the Earth* (RE), part 3, "Catholic Social Teaching and Environmental Ethics," section A, "A Sacramental Universe."

universe grounds human accountability for the fate of the earth."[38] They assert *that* articulation of the seventh theme (not the usual "care for God's creation") drawn from the tradition of Catholic social teaching offers "a developing and distinctive perspective on environmental issues."[39] In *Renewing the Earth* the bishops stress the reality that the earth is God's creation, that there is a divine revelation that comes to us through creation, and that it is thus "sacramental"—more than just "stuff" like the things humans create to be used indiscriminately.

In a dialogical tone, the bishops acknowledge: "Although Catholic social teaching does not offer a complete environmental ethic, we are confident that this developing tradition can serve as the basis for Catholic engagement and dialogue with science, the environmental movement, and other communities of faith and good will."[40]

The bishops reflect on primordial and human history, which provides evidence that people have continually perceived the sacred in creation. This evidence is immediately contextualized within the Christian narrative: "The Christian vision of a sacramental universe —a world that discloses the Creator's presence by visible and tangible signs—can contribute to making the earth a home for the human family once again."[41]

The bishops again assert a *possibility* when they state: "Reverence for the Creator present and active in nature, moreover, *may* serve as ground for environmental responsibility."[42] This stance is sustained by the claim that humans are related to creation but part of the natural world: "Safeguarding creation requires us to live responsibly within it, rather than manage creation as though we are outside it."[43]

The bishops make an important distinction between the claim that creation is a gift of God, the capacity of the universe to disclose something of God, and the human capacity to receive and respond to that disclosure: "The Christian vision of a sacramental universe—a world that discloses the

38. RE, part 3, "Catholic Social Teaching and Environmental Ethics," section A.

39. RE, part 3, "Catholic Social Teaching and Environmental Ethics," section A; see also the USCCB's list of the Seven Principles of Catholic Social Teaching at http://www.usccb.org/beliefs-and-teachings/what-we-believe/catholic-social-teaching/seven-themes-of-catholic-social-teaching.cfm.

40. RE, part 2.

41. RE, part 3, A.

42. RE, part 3, A; emphasis added.

43. RE, part 2, A.

Creator's presence by visible and tangible signs—*can contribute* to making the earth a home for the human family once again."[44]

Sacramentality and Ritual Prayer

For Catholic environmental ethics and liturgical pedagogy, it is vital to retain the distinction between a creation theology and sacramentality. "The world itself is not a sacrament, but it is a place in which sacramentality is possible."[45] Sacramentality is always a particularized event that involves God's disclosure and offer of grace, but it is also a human response to the divine. All of this takes place in a context that is the created world itself. The liturgy is pedagogical (formative or catechetical) when it supplies "not the answers, not the concrete socioeconomic steps today . . . but rather the sanction, inspiration, orientation, and dynamic to move believers to join with other people and groups who want to make the world a better home for all. Our rites do not excuse us from work, they impel us to work."[46]

Ritual sacramentality and the sacramentality of the world are related; indeed, a sacramental worldview can be a motive for environmental ethical action. An occasion for liturgical pedagogy exists in the negative contrast experience available in ritual worship, motivating ethical action. One's perception of a sacramental world is largely dependent on whether or not one views the world as *creation*, that is, as a gift.[47] According to theologian Louis-Marie Chauvet, receiving the world as gift implies the necessity of a return-gift. In this light, creation is shot through with sacramentality. For Christians, it is *in* the mystery of the liturgy and ritual action that faith confesses that *God is the Creator*.[48] It is also *through* seeing the sacramental structure of creation that it is possible to apprehend that all reality is integral to the history of salvation, which begins with creation.

Kevin Irwin explains that "belief in the God of creation and redemption is helpfully reflected in both the fact that we use symbols in the liturgy but also in ways which they are used," for example, for washing, eating and

44. RE, part 3, A; emphasis added.

45. Osbourne, *Christian Sacraments*, 140.

46. Robert W. Hovda, "The Relevance of the Liturgy," *Worship* 64, no. 5 (September 1990): 449–50.

47. Louis-Marie Chauvet, *Symbol and Sacrament: A Sacramental Reinterpretation of Christian Existence*, trans. Patrick P. Madigan and Madeleine E. Beaumont (Collegeville, MN: Liturgical Press, 1995), 228.

48. Chauvet, *Symbol and Sacrament*, 555.

sharing, anointing, etc. At times, it is the very use of the symbols that vividly raises up the need for critical care for the planet "because what we do liturgically clashes with what we now find in nature and in the cosmos."[49] How can one possibly think of "cleansing" when the only water one can access is polluted and not drinkable? Or when so little water is available that life itself is threatened? Or when not enough food for proper basic nutrition is available at a banquet?

This reality is less a matter of injecting a political agenda than simply acknowledging the clash between what we profess and what the actuality of life demonstrates as the current environmental and ecological status of the earth.[50]

Belgian Cardinal Godfried Danneels argues that it is "part of human nature to engage in liturgical practices, and thus people have an innate capacity and proclivity to be open, receptive, ready to listen, giving and receiving in relation to the created world."[51] Feminist theologian, Susan Ross rightly asserts, "The sacramental principle is not an invention of Christianity. . . . Sacramentality is not simply a general principle, but is itself constitutive of revelation. . . . There is not only a logic and order to creation (natural law) but a revelatory dimension to nature that goes beyond logic to an encounter with the source of life."[52]

Beyond this, any adequate approach to teaching the relationship between sacraments and ethics should include the practice of justice, the moral power of symbols and rituals, and the need to extend sacramentality beyond the "seven sacraments."[53] There is an urgent need today for those who write, teach, preach, and promulgate Catholic social teaching to mind creditably the gap between ethics and ritual. Pope Francis has called all Catholics to that continuity in worship and ethical praxis.

---

49. Kevin W. Irwin, "Liturgical *Actio*: Sacramentality, Eschatology And Ecology," *Questions liturgiques* 81, nos. 3–4 (2000): 121.

50. See Irwin, "Liturgical Actio," 121–22.

51. Jame Schaefer, *Theological Foundations for Environmental Ethics: Reconstructing Patristic and Medieval Concepts* (Washington, DC: Georgetown University Press, 2009), 87. She cites John Hart, Sacramental Commons: Christian Environmental Ethics (Lanham, MD: Rowman & Littlefield, 2006), xiii. See Danneels, "Liturgy Forty Years after the Second Vatican Council: High Point or Recession," in *Liturgy in a Postmodern World*, ed. Keith Pecklers (New York: Continuum, 2003), 10.

52. Susan Ross, *Extravagant Affections: A Feminist Sacramental Theology* (New York: Continuum, 2001), 35.

53. Ross, *Extravagant Affections*, 179.

World Day of Prayer for the Care of Creation: An Opportunity for Liturgical Catechesis

On August 6, 2015, Pope Francis instituted September 1 as the annual World Day of Prayer for the Care of Creation. On this day, Catholics are to honor creation jointly with Orthodox Christians. He holds that particularly Orthodox Christians (who emphasize the sacramentality of creation) can help Western Christians "rediscover in our rich spiritual patrimony the deepest motivations for our concern for the care of creation."[54] Our acts of prayer on this day should thus link worship to ethical action. The annual World Day of Prayer for the Care of Creation brings to light what the Church does at each liturgy through the celebration of Word and Sacrament. This could take a variety of forms:

- Pray at mealtimes. Before and after meals, say a short prayer of thanksgiving for the life-giving food that sustains and nourishes us.[55]
- Counteract the "throwaway culture." In *Laudato si'*, Pope Francis brings attention to our "throwaway culture," which "quickly reduces things to rubbish."[56] In your daily life, try to identify the ways in which you can choose reusables rather than disposables, such as coffee mugs, reusable bags, or cloth napkins, and commit to making one change during this month.
- Bring your "ecological offences" to the sacrament of Reconciliation. In calling for a deep "ecological conversion," Pope Francis has emphasized the importance of examining one's own conscience, of recognizing one's sins against creation, however great or small. Seeing the interconnectedness of our world leads to an understanding that "[e]very violation of solidarity and civic friendship harms the environment."[57] Bring these sins to the sacrament of Reconciliation and perform a spiritual work of mercy for

54. "Letter of His Holiness Pope Francis for the Establishment of the World Day of Prayer for the Care of Creation," September 1, 2018; available from http://w2.vatican.va/content/francesco/en/messages/pont-messages/2018/documents/papa-francesco_20180901_messaggio-giornata-cura-creato.html.

55. See LS, 227.

56. LS, 22.

57. CV, 51.

our common home, such as an act of "grateful contemplation of God's world."[58]

- Organize an educational program in your parish or with a group of friends. Use the educational program "Befriend the Wolf" from the Catholic Climate Covenant to reflect on our vocation as stewards of creation. The program is designed to help your community contemplate the connections between all creatures under God our Creator.[59]

As Pope Francis writes:

> The Sacraments are a privileged way in which nature is taken up by God to become a means of mediating supernatural life. Through our worship of God, we are invited to embrace the world on a different plane. Water, oil, fire and colors are taken up in all their symbolic power and incorporated in our act of praise. The hand that blesses is an instrument of God's love and a reflection of the closeness of Jesus Christ, who came to accompany us on the journey of life. Water poured over the body of a child in Baptism is a sign of new life. Encountering God does not mean fleeing from this world or turning our back on nature. This is especially clear in the spirituality of the Christian East. Beauty, which in the East is one of the best loved names expressing the divine harmony and the model of humanity transfigured, appears everywhere: in the shape of a church, in the sounds, in the colors, in the lights, in the scents.[60]

For Christians, all the creatures of the material universe find their true meaning in the incarnate Word, for the Son of God has incorporated in his person part of the material world, planting in it a seed of definitive transformation. "Christianity does not reject matter. Rather, bodiliness is considered in all its value in the liturgical act, whereby the human body is disclosed in its inner nature as a temple of the Holy Spirit and is united with the Lord Jesus, who himself took a body for the world's salvation."[61] Ultimately, when consistent and explicit Christian ethical praxis bears witness to the vital coherence between *lex orandi, lex credendi,* and *lex vivendi,* and *only* then, will the Church recognize the sacramentality of God's creation. For, "when we can see God

58. LS, 214.

59. You can access this resource by visiting https://catholicclimatecovenant.org/program/feast-st-francis.

60. LS, 235.

61. LS, 235.

reflected in all that exists, our hearts are moved to praise the Lord for all his creatures and to worship him in union with them."[62]

Questions for Parish Staff Discussion and Reflection

- Does the environment in the worship space and on the church grounds reflect the parish's care for creation? Do the Greeting and Usher ministries see themselves as creators and stewards of an environment of welcome and hospitality for all who enter the sacred space?
- Do the prayers that are written and selected as well as the homilies include connections to the sacredness of the liturgy, the sacred space, and the creation of God's world?
- Does the music ministry incorporate hymns as appropriate that reflect our regard and reverence for God's creation, for both our "Sister, Mother Earth," and human kind?[63]
- Is a culture nurtured within the assembly such that all who are gathered are aware that they have a role in helping to create and complete the environment for the praise and worship of God?
- Do homilies speak of our Christian belief that all creation is sacred and highlight the violations of governments, corporations, and other groups who abuse, rape, and desecrate the earth and its resources, causing further harm in places where there is already extreme poverty, thus destroying the posterity of the earth's resources for future generations?
- Does the parish encourage simple living, care for the environment, recycling, composting, and the use of energy saving and renewable energy sources?
- Does the peace and justice/faith-in-action committee offer education on environmental concerns and our personal and corporate responsibility to care for the environment and teach our children to do the same? Do they highlight local abuses of the environment through pollution of the air, water, or aquifers or dumping of toxic substances in poor communities?[64]

---

62. LS, 87.

63. St. Francis of Assisi as quoted in *Laudato si'*, 1.

64. Questions for discussion and reflection for young adults may be downloaded for free at https://ltp.org/LCST.

APPENDIX

Penitential Services

Written and compiled by Larry Dowling

IN ITS APPENDIX, THE *RITE OF PENANCE* INCLUDES SEVERAL MODELS of nonsacramental penitential services. These celebrations provide a number of options based on the liturgical season, on particular themes, and on the makeup of a particular group. The services may be adapted to pastoral circumstances, keeping in mind the specific conditions and needs of the particular community that gathers for the celebration. Using these services as a model, this appendix includes seven nonsacramental penitential services, each based on one of the seven themes of Catholic social teaching.

Offering and praying these services within the parish will help open the hearts of our assembly members to be transformed by the love and mercy of Christ and to go forth as his disciples and heal the brokenness in our world today. They may take place at any time throughout the liturgical year. These services may also be downloaded as a free PDF on this website: https://ltp.org/LCST. These services may be reproduced only in quantities necessary for the church, school, or group purchasing this resource. The PDFs must be reproduced with their accompanying copyright notices. Reproduction of any other part of this resource for any other purpose is both illegal and unethical.

THEME 1
The Life and Dignity of the Human Person

All stand and sing the opening song.

Opening Song

A song from the following options may be chosen or you may select another song that is more familiar to your community.

- "Gather Us In" by Marty Haugen (GIA; sing verses 1, 2, and 4)
- "Christ, Be Our Light" by Bernadette Farrell (OCP)
- "The Servant Song" by Richard Gillard (various publishers)
- "Canticle of the Turning" by Rory Cooney (GIA)
- "All Are Welcome" by Marty Haugen (GIA)

After the opening song, the presider and the faithful sign themselves with the Cross. The minister then offers the greeting.

Presider: In the name of the Father, and of the Son, and of the Holy Spirit.

All: **Amen.**

Presider: The Lord be with you.

All: **And with your spirit.**

If the presider is a lay minister, use the following greeting and response:

Presider: Let us praise our Lord Jesus Christ, who loved us and gave himself for us. Let us bless him now and forever. Blessed be God for ever.

All: **Blessed be God for ever.**

After an appropriate song and the greeting by the minister, the meaning of this penitential service is explained to the people. The minister then leads this prayer, which helps the Christian faithful call to mind those times that they did not respect the fundamental dignity given to all human persons.

Presider: My brothers and sisters, there are times in which we have neglected the sacredness of life and the dignity of the human person. Let us ask God to renew his grace within us as we turn to him in repentance, seeking to change our actions, respecting and loving all peoples in this world.

Presider: Let us kneel (**or:** Bow your heads before God).

All pray in silence for a brief period.

Presider: Let us stand (**or:** Raise your heads).

Presider: Almighty and Ever Life-Giving God,
you sent Your Only Beloved Son, the Word Made Flesh,
among us,
to bless our common humanity
and bond us eternally in Spirit with you.
May we always reverence this precious gift
in all of our brothers and sisters.
Through our Lord Jesus Christ, your Son,
who lives and reigns with you in the unity
of the Holy Spirit,
one God, for ever and ever.

All: **Amen.**

All are seated for the Liturgy of the Word. The following readings with corresponding Lectionary numbers are provided as options. Other readings may also be chosen. Parish readers and cantors may be involved.

First Reading

- Genesis 1:26—2:3 (LM, #559; Proper of Saints)
- Deuteronomy 30:15–20 (LM, #220; Weekday)
- 1 Corinthians 6:13c–15a, 17–20 (LM, #65B)
- 1 Corinthians 3:9c–11, 16–17 (LM, #671; Proper of Saints)
- 1 John 3:1–2 (LM, #50B)

Responsorial Psalm

Sing a setting of one of the following psalms that is familiar to your parish.

- Psalm 8:4–5, 6–7, 8–9 (LM, #166C)
- Psalm 22:8–9, 17–18, 19–20, 23–24 (LM, #38ABC)
- Psalm 25:4–5, 6–7, 8–9 (LM, #23B)
- Psalm 34:2–3, 4–5, 6–7, 8–9, 10–11 (LM, #978, 2; Votive Mass for the Holy Eucharist)
- Psalm 139:1–3, 7–8, 9–10, 13–14ab (LM, #459; Weekday, Year II)

Gospel Acclamation

All stand for the Gospel acclamation and the proclamation of the Gospel. Sing a setting of the acclamation that is familiar to your parish. Use the alternate acclamations during Lent.

Gospel

- Matthew 19:13–15 (LM, #418; Weekday Year II)
- Matthew 11:25–30 (LM, #100A)
- Matthew 25:31–46 (LM, #160A)
- Luke 10:25–37 (LM, #105C)
- John 3:14–21 (LM, #32B)

Homily/Reflection

After the proclamation, all are seated for the homily or reflection.

Preaching Points

The presider may use the following points to prepare a homily or reflection that is focused on human dignity.

- The dignity and sanctity of human life is at the core of Catholic social teaching. It undergirds every principle. Grounded "in the beginning" with the Genesis story and confirmed and blessed through the Incarnation, each human person has the potential to reveal themselves as God-like. Life is sacred and must be treated with dignity from the moment of conception through the moment of death—at every age and every stage of life.
- There are many social threats to human dignity: abortion and child abuse; domestic abuse and human trafficking; racism and sexism;

discrimination against migrants, refugees, and immigrants; the failure to provide adequate health care and livable and affordable housing; elder abuse; and many other sins against human life. In both the Old and New Testaments, the Scripture accounts are very clear that the love of God is equated to how we love ourselves and our neighbors. We must seek to end all threats to human dignity.

- Jesus' ministry centered on healing and the casting out of demons from those who were clearly caught up in self-destructive behavior. As a people of faith, we are called within family, our faith community, and beyond to fight for the dignity of every person, each of us precious and sacred in the eyes of God.

Examination of Conscience

After the homily or reflection, the examination of conscience occurs. Provide the assembly with time of silence to personally reflect after each examen. The assembly may kneel or stand. The examen should be read aloud by a reader.

Presider: St. Paul says: "Do you not know that you are God's temple and that God's Spirit dwells in you? If anyone destroys God's temple, God will destroy that person. For God's temple is holy, and you are that temple."[1]

Reader: Do I respect myself as a temple of God's Spirit? Do I care for my body as I should? Do I eat properly and exercise as I should? Do I exercise my mind, opening myself to new learning and growth? Do I care for my mind and soul through daily prayer, reading of the Scriptures, and connecting with others who support and challenge me?

Pause for a moment of silent prayer.

Reader: Do I treat my spouse, my friends, my coworkers, and/or my classmates with dignity, love, and respect? Do I help nurture their gifts and encourage the sharing of these gifts?

Pause for a moment of silent prayer.

Reader: Have I ever objectified another person as a source of pleasure? Have I viewed pornographic images that demean and objectify the bodies of men or women?

Pause for a moment of silent prayer.

1. 1 Corinthians 3:16–17.

Reader: Have I treated women with the respect they deserve? Have I been disrespectful, even abusive, in words and actions, against a spouse or friend?

Pause for a moment of silent prayer.

Reader: Have I encouraged others to properly care for themselves and develop their talents and gifts?

Pause for a moment of silent prayer.

Reader: Have I expressed racist or prejudiced attitudes toward people of different races, ethnicities, or religious beliefs?

Pause for a moment of silent prayer.

Reader: Do I support, to the best of my ability, women and girls who are struggling with decisions related to pregnancy? Do I encourage and accompany them from pregnancy through birth and beyond?

Pause for a moment of silent prayer.

Reader: Have I damaged others through gossip, defamation of character, false judgments, or theft?

Pause for a moment of silent prayer.

Reader: Have I caused physical or emotional harm to others?

Pause for a moment of silent prayer.

Reader: Do I support efforts to protect the unborn, the homeless, victims of domestic violence, victims of child abuse, and the elderly?

Pause for a moment of silent prayer.

Reader: Do I teach my children to respect all people regardless of race, ethnicity, gender, faith tradition, or sexual orientation?

Pause for a moment of silent prayer.

Reader: Do I love my enemies to the extent that I will, at a minimum, pray for them and, at my best, seek to build bridges of reconciliation and understanding?

Pause for a moment of silent prayer.

Act of Repentance

After the examination of conscience, the presider invites all to prayer in these or similar words.

Presider: My brothers and sisters, in God we are "fearfully and wonderfully made."[2] Because God's works are wonderful, we give him praise.[3] Humbly and confidently, let us ask God to renew us in body, mind, and spirit. May our lives reflect the passionate love of our Creator and conform more fully to the ways of Christ.

Reader: "Lord, when did we see you hungry and not feed you?" All too often, we have fed the unhealthy hungers in ourselves, denigrating our bodies, our minds, and our spirits. We have failed to respond to the legitimate hungers of our family members for attention, affirmation, presence, and love. Lord, we seek forgiveness and long to know the Master's joy:

R: **Have mercy on us!**

Reader: "Lord, when did we see you thirsty and not give you drink?" At times we thirst for attention, for superiority, and for material and emotional wants that we really do not need or are unhealthy for us. At times, we fail to respond to the poor who are thirsting for simple recognition and assistance. Lord, we seek forgiveness and long to know the Master's joy:

R: **Have mercy on us!**

Reader: "Lord, when did we see you a stranger and not welcome you?" At times we buy into stereotypes about people of different races, ethnicities, economic and social classes, religions, or sexual orientation. We fail to take the risk of getting to know someone who is different from us and letting them get to know us, thus discovering that we are not really strangers. Lord, we seek forgiveness and long to know the Master's joy:

R: **Have mercy on us!**

2. Psalm 139:14.
3. See Psalm 139:14.

Reader: "Lord, when did we see you naked and not clothe you?" There have been times in our life when we may have faced physical or verbal abuse, an addiction, or other great challenges that left us feeling naked, hopeless, and alone. Yet now that we have been healed, we are resistant to sharing our stories with others facing similar challenges. These people need to hear our stories of hope and of how we overcame crisis and trauma; yet we close our eyes to the plight of these people who are stripped of their dignity and respect. Lord, we seek forgiveness and long to know the Master's joy:

R: **Have mercy on us!**

Reader: "Lord, when did we see you ill and not minister to you?" At times we fail to recognize our own illnesses, our own insecurities, and are afraid to reach out to others, feeling they will think that we are weak. We also fail at times to recognize the needs of those whose bodies, minds, and spirits are suffering: the mentally ill, the abused elderly, the physically and verbally abused spouse or child, or the pregnant woman sick worried and fearful about how she will be able to care for her child. Lord, we seek forgiveness and long to know the Master's joy:

R: **Have mercy on us!**

Reader: "Lord, when did we see you in prison and not visit you?" We fail to see that many members of our society are imprisoned in poverty because of poor education. Or those who have criminal records because of traumas from war, famine, natural disasters, and individual and corporate greed. We resist doing anything to remove the bars that keep them from pursuing their God-given potential. Lord, we seek forgiveness and long to know the Master's joy:

R: **Have mercy on us!**

Reader: "Lord, when did we not see you in our brothers and sisters?" We all too often disregard the casualties of war throughout our world, the genocide in our own cities, and the collateral damage of men, women, and children. We may perceive others as "less than" or "expendable" because of distance, poverty, religious belief, or the color of their skin. We may fail to seek ways to defend their dignity and to demand that those in power regard them as our brothers and sisters, equal in the eyes of God. Lord, we seek forgiveness and long to know the Master's joy:

R: **Have mercy on us!**

If the assembly has been kneeling for the examination of conscience and act of repentance, they may stand at this time for the remainder of the service.[4]

Presider: Through the Lord's Incarnation, his coming as man, the Word Made Flesh, God sanctified humanity and dignified us with his grace. Let us pray in the words his Son gave us.

Our Father . . .

Presider: Lord, you have blessed us with the precious gift of life,
a gift ever to be reverenced, nurtured, and shared.
We thank you and pledge our very lives
to uphold the dignity and sanctity of all life,
by binding ourselves ever to you
who are the source of life and the font of love eternal.
Through Christ our Lord.

All: **Amen.**

Presider: The Lord be with you.

All: **And with your spirit.**

Presider: May almighty God bless you,
the Father, and the Son, and the Holy Spirit.

All: **Amen.**

4. The act of repentance is based on Matthew 25:44.

If a lay minister is the presider, he or she uses this final blessing:

Presider: May the Lord bless us,
protect us from all evil,
and bring us to everlasting life.

All: **Amen.**

Presider: Go in peace, glorifying the Lord by your life.

All: **Thanks be to God.**[5]

Song of Mission

The service concludes with a rousing song commissioning the assembly to go forth in peace and love. Other more familiar songs may also be chosen.

- "We Are Called" by David Haas (GIA)
- "Lord, Whose Love Through Humble Service" (traditional; various publishers)
- "Make Me a Channel of Your Peace" (traditional; various publishers)
- "Voices That Challenge" by David Haas (GIA)
- "We Are the Light of the World" by Jean A. Grief (Vernacular Hymns Publishing Co.)

5. The greeting, responses, and final dismissal are based in part on those found in *The Roman Missal*, *The Liturgy of the Hours*, and the *Book of Blessings*, International Commission on English in the Liturgy.

THEME 2
The Call to Family, Community, and Participation

All stand and sing the opening song.

Opening Song

A song from the following options may be chosen or you may select another song that is more familiar to your community.

- "We Are Many Parts" by Marty Haugen (GIA)
- "Song of Gathering" by Joe Wise (GIA)
- "God Is Here! As We His People" (Hope Publishing; traditional)
- "No Longer Strangers" by David Haas (GIA)
- "In Christ There Is No East or West" (traditional)

After the opening song, the presider and the faithful sign themselves with the Cross. The minister then offers the greeting.

Presider: In the name of the Father, and of the Son,
and of the Holy Spirit.

All: **Amen.**

Presider: The Lord be with you.

All: **And with your spirit.**

If the presider is a lay minister, use the following greeting and response:

Presider: The grace of our Lord Jesus Christ be with us all,
now and for ever.

All: **Amen.**

After an appropriate song and the greeting by the minister, the meaning of this penitential service is explained to the people. The minister then leads this prayer, which helps the Christian faithful call to mind those times that they did not respect the bonds of community as a living, efficacious sign of divine love in the world.

Presider: Fruitful love is an icon of the living God. Yet there are times when we have violated, abused, and rejected this sacred bond. Let us ask God to renew his grace within us as we turn to him in repentance. "May we never stop seeking that fullness of love and communion which God holds out before us."[1]

Presider: Let us kneel (**or:** Bow your heads before God).

All pray in silence for a brief period.

Presider: Let us stand (**or:** Raise your heads).

Presider: Gracious God, source of unity and love,
you bind us together as one family
through the love of the Most Holy Trinity:
Father, Son, and Spirit.
Help us to support every family, in whatever shape or form,
that more and more we may become the
beloved community
for which you long for us to become.
Through Christ our Lord.

All: Amen.

All are seated for the Liturgy of the Word. The following readings with corresponding Lectionary numbers are provided as options. Other readings may also be chosen. Parish readers and cantors may be involved.

First Reading

- Genesis 4:1–15, 25 (LM, #335; Weekday, Year I)
- Jeremiah 7:1–11 (LM, #401; Weekday, Year II)
- James 2:14–18 (LM, #131B)
- 1 Peter 4:7–13 (LM, #351; Weekday, Year II)
- 1 John 4:19—5:4 (LM, #215; Weekday)

Responsorial Psalm

Sing a setting of one of the following psalms that is familiar to your parish.

- Psalm 34:2–3, 4–5, 6–7, 8–9, 17–18, 19, 23 (LM, #997)
- Psalm 72:1–2, 7–8, 12–13, 17 (LM, #4A)

1. *Amoris laetitia*, 325.

- Psalm 85:7–8, 9, 11–12 (LM, #954.I; Masses for Various Needs and Occasions)
- Psalm 100:1b–2, 3, 4, 5 (LM, #208; Weekday)
- Psalm 103:1–2, 3–4, 8–9, 11–12 (LM, #960.2; Masses for Various Needs and Occasions)

Gospel Acclamation

All stand for the Gospel acclamation and the proclamation of the Gospel. Sing a setting of the acclamation that is familiar to your parish. Use the alternate acclamations during Lent.

Gospel

- Matthew 12:46–50 (LM, #396; Weekday)
- Matthew 22:34–40 (LM, #148A)
- Matthew 5:13–16 (LM, #73A)
- John 15:1–8 (LM, #53B)
- John 15:12–17 (LM, #289; Weekday)

Homily/Reflection

After the proclamation, all are seated for the homily or reflection.

Preaching Points

The presider may use the following points to prepare a homily or reflection that is focused on the call to family, community, and participation.

- The Church teaches that the family is the core institution of every society. The family is "also be a sign of unity for the world by presenting to their children a model of life based on the values of truth, freedom, justice, and love."[2] The family unit is where our children learn compassion and forgiveness and how to communicate in positive ways, to peacefully resolve conflict, to work together, and to contribute in mutual care to its own growth. The family unit also provides an environment in which love is experienced and finds expression in the positive acts of charity and justice—within and beyond the home (community).
- Scripture texts continually highlight the importance of family. In the Book of Genesis, God creates the first couple, Adam and Eve,

2. *The Family in the Modern World*, 48.

and the first family (siblings Cain and Abel). The first sin after the fall was the murder of a sibling prompted by jealousy! Throughout the Old Testament, we have examples of imperfect families all seeking in some way to be faithful to God. The sins against family are often internal, in the ways family members treat each other—sometimes by abuse, neglect, infidelity, selfishness, and addiction.

- There are many external forces that negatively affect families: social and business pressures; unemployment, overemployment, or underemployment; unlivable or unaffordable housing; poor educational institutions in communities; violence; drugs; alcohol; lack of potable water; lack of access to healthy foods and good health care; anti-immigrant policies.
- As Christians, we are called to look within our hearts to determine what we must do to help strengthen our own families in faith, mutual nurturance, and growth. As part of various communities (faith communities, social communities), God calls us to assist other families in need—providing the basic needs of food, clothes, and shelter, or assistance in addressing issues that are causing pain and division in homes. There are wonderful stories in the Scriptures about how Jesus brought healing to families; for example, the daughter of Jairus, the raising of the son of the widow of Nain, and the raising of Lazarus. Of course, where we are able, God invites our families to extend a hand in helping families beyond our communities.

Examination of Conscience

After the homily or reflection, the examination of conscience occurs. Provide the assembly with a time of silence to personally reflect on each examen. The assembly may kneel or stand. The examen should be read aloud by a reader.

Presider: Jesus said: "This is my commandment: love one another as I have loved you."[3] Therefore, in order to love God, we must also love our brothers and sisters.[4]

Reader: Have I done my best as a parent for my children? Have I raised them with Christian values, forming them in faith, and encouraging and lovingly challenging them to grow into men and women of integrity and generosity? Have I exercised discipline in ways that are constructive and

3. John 15:12–17.
4. See 1 John 4:19–21.

conducive to growth and learning? Have I encouraged unity and harmony within my family?

Pause for a moment of silent prayer.

Reader: Have I ever done physical or emotional harm to my spouse or children?

Pause for a moment of silent prayer.

Reader: Am I able to forgive and accept forgiveness from other family members?

Pause for a moment of silent prayer.

Reader: Do I love my neighbors? Do I have a genuine care and concern for their well-being and prosperity?

Pause for a moment of silent prayer.

Reader: Have I been true to all of the relationships that God has blessed me with in my life?

Pause for a moment of silent prayer.

Reader: Do I respect people of different races, ethnicities, religions, and sexual orientations? Have I ever participated in demeaning or denigrating anyone because they are "different"?

Pause for a moment of silent prayer.

Reader: Do I support my Church to my fullest, sharing my time, talent, and treasure for the collective family of the Body of Christ in my parish?

Pause for a moment of silent prayer.

Reader: Do I participate in activities that support and help strengthen other families in my community and beyond?

Pause for a moment of silent prayer.

Reader: Do I respond to appeals that support individuals and families suffering in the midst of war, famine, and natural disasters?

Pause for a moment of silent prayer.

Reader: Have I responded to the best of my ability to the needs of my neighbors, the homeless, the poor, and the imprisoned?

Pause for a moment of silent prayer.

Act of Repentance

After the examination of conscience, the presider invites all to pray in these or similar words:

Reader: My brothers and sisters, "How wonderful and pleasant it is when brothers and sisters live together in harmony."[5] With sincere hearts, let us ask God to remind us that we are all members of a family and are collectively bonded in the Spirit as one human family. May God's grace open our hearts and minds, that we may solidify those bonds that at times have been tested by our own actions or inactions.

Reader: "Lord, when did we see you hungry and not feed you?" We fail at times to recognize the hunger in our own families: hunger for support and recognition, for a sense of belonging, for a desire for unity and harmony in our homes. We falter in our response to provide for the basic needs of families in our community and beyond. Lord, we seek forgiveness and long to know the Master's joy:

R: Have mercy on us!

Reader: "Lord, when did we see you thirsty and not give you drink?" We fail to recognize that many in our communities thirst for clean water and the necessities of life that we may take for granted. They thirst for us to respond to their plight. They thirst for us to reach out to renew their parched bodies and spirits, to give them opportunities to live life to its fullest and not just try to survive the trials of each day. Lord, we seek forgiveness and long to know the Master's joy:

R: Have mercy on us!

5. Psalm 139:14.

Reader: "When did we see you a stranger and not welcome you?" We are aware of so many people in our world who are migrants, refugees, and immigrants. They are fleeing economic hardship, political oppression, and violence. They are seeking freedom from oppression and asylum in a new home. We don't welcome them. We don't treat them as brothers and sisters in need. We divide their humanity with borders. It is our responsibility—and within our abilities—to welcome them and help them find the stability we often take for granted. Lord, we seek forgiveness and long to know the Master's joy:

R: **Have mercy on us!**

Reader: "When did we see you naked and not clothe you?" We fail to recognize the many members of the human family, even in our own community and nation, who have been stripped of their basic rights because of race, ethnicity, religious belief, or sexual orientation. We hold our own prejudices and do little to clothe ourselves and our communities in awareness, education, and dialogue and encourage tolerance and hospitality. Lord, we seek forgiveness and long to know the Master's joy:

R: **Have mercy on us!**

Reader: "When did we see you ill and not minister to you?" We fail at times to acknowledge the real illnesses that affect whole communities in our country and in our world: diseases caused by pollution, corporate and governmental abuses of the environment, and the lack of immediate response by the world community to natural disasters. We fail to address and seek healing for the illness of apathy and indifference in ourselves and in our parish community. Lord, we seek forgiveness and long to know the Master's joy:

R: **Have mercy on us!**

Reader: "When did we see you in prison and not visit you?" We resist addressing the imprisonment of family members in our community affected by domestic violence; the physical, verbal, and sexual abuse of children by adults; illiteracy, addiction, mental illness, and trauma. We fail to engage our communities in responding to these problems prevalent in our community. Problems that imprison our brothers and sisters behind the walls of their homes and within walls they have created within themselves to cope with the hurt. Lord, we seek forgiveness and long to know the Master's joy:

R: Have mercy on us!

Reader: "Lord, when did we not see you in our brothers and sisters?" We struggle at times to see certain people as part of the human family, as brothers and sisters in faith, and as equal and precious in God's eyes. We are blinded by prejudice, and we fail to address this blindness within ourselves and within in our community. Lord, we seek forgiveness and long to know the Master's joy:[6]

R: Have mercy on us!

If the assembly has been kneeling for the examination of conscience and act of repentance, they may stand at this time for the remainder of the service.

Presider: Through Christ, we become adopted sons and daughters of God. As one in faith, we have the courage to pray in the words he gave us:

Our Father . . .

6. The act of repentance is based on Matthew 25:44.

Presider: Jesus, our brother and Lord,
Son of God and Son of Mary,
in Word and deed you sought to heal and reconcile
relationships within families and communities.
May we claim our dignity as part of this
great human family
and work to protect and fortify every family
that communities might be strengthened,
anchored in true justice and lasting peace.
Who live and reign with God the Father in the
unity of the Holy Spirit,
one God, for ever and ever.

All: **Amen.**

Presider: The Lord be with you.

All: **And with your spirit.**

Presider: May almighty God bless you,
the Father, and the Son, and the Holy Spirit.

All: **Amen.**

If a lay minister is the presider, he or she uses this final blessing:

Presider: May the Lord bless us,
protect us from all evil,
and bring us to everlasting life.

All: **Amen.**

Presider: Go in peace, glorifying the Lord by your life.

All: **Thanks be to God.**[7]

7. The greeting, responses, and final dismissal are based in part on those found in *The Roman Missal*, *The Liturgy of the Hours*, and the *Book of Blessings*, International Commission on English in the Liturgy.

Song of Mission

The service concludes with a rousing song commissioning the assembly to go forth in peace and love. Other more familiar songs may also be chosen.

- "We Are Called" by David Haas (GIA)
- "Prayer of Peace" by David Hass; based on a Navajo prayer (GIA)
- "City of God" by Daniel L. Schutte (New Dawn Music)
- "Voices That Challenge" by David Haas (GIA)
- "Lord, Whose Love in Humble Service" (traditional; various publishers)

THEME 3
Rights and Responsibilities

All stand and sing the opening song.

Opening Song

A song from the following options may be chosen or you may select another song that is more familiar to your community.

- "There's a Wideness in God's Mercy" (traditional; various publishers)
- "Here I Am, Lord" by Daniel L. Schutte (New Dawn Music)
- "We Will Serve the Lord" by Rory Cooney (North American Liturgy Resources)
- "Christ Be Our Light" by Bernadette Farrell (OCP)
- "Whatsoever You Do" by William F. Jabusch (GIA)

After the opening song, the presider and the faithful sign themselves with the Cross. The minister then offers the greeting.

Presider: In the name of the Father, and of the Son, and of the Holy Spirit.

All: **Amen.**

Presider: The Lord be with you.

All: **And with your spirit.**

If the presider is a lay minister, use the following greeting and response:

Presider: Let us praise God our creator, who gives us courage and strength. Blessed be God for ever.

All: **Blessed be God for ever.**

After an appropriate song and greeting by the minister, the meaning of this penitential service is explained to the people. The minister then leads this prayer, which helps the Christian faithful call to mind those times that they did not respect the fundamental rights and responsibilities of the people of God.

Presider: My brothers and sisters, there are times in which we have neglected the fundamental rights and responsibilities that God has given all peoples. Let us ask God to renew his grace within us as we turn to him in repentance, seeking to change our actions, respecting and loving all peoples in this world.

Presider: Let us kneel (**or:** Bow your heads before God).

All pray in silence for a brief period.

Presider: Let us stand (**or:** Raise your heads).

Presider: God our Father,
source of life and font of love,
you remind us that true freedom
finds its origin and fulfillment in you.
Help us claim with humility
both the privilege and responsibility of serving others,
that all might know true freedom and newness of life.
Through Christ our Lord.

All: **Amen.**

All are seated for the Liturgy of the Word. The following readings with corresponding Lectionary numbers are provided as options. Other readings may also be chosen. Parish readers and cantors may be involved.

First Reading

- Leviticus 19:1–2, 17–18 (LM, #79A)
- Acts of the Apostles 4:32–35 (LM, #44B)
- 1 Corinthians 3:9c–11, 16–17 (LM, #671; Proper of Saints)
- 2 Corinthians 9:6–11 (LM, #367; Weekday, Year I)
- James 2:14–18 (LM, #131B)

Responsorial Psalm

Sing a s setting of one of the following psalms that is familiar to your parish.

- Psalm 8:4–5, 6–7, 8–9 (LM, #166C)
- Psalm 107:2–3, 4–5, 6–7, 8–9 (LM, #423; Weekday, Year II)
- Psalm 112:1bc–2, 3–4, 5–6, 7–8, 9 (LM, #623; Proper of Saints)

- Psalm 119:23–24, 26–27, 29–30 (LM, #273; Weekday, Year I)
- Psalm 119:29, 43, 79, 80, 95, 102 (LM, #407; Weekday, Year II)

Gospel Acclamation

All stand for the Gospel acclamation and the proclamation of the Gospel. Sing a setting of the acclamation that is familiar to your parish. Use the alternate acclamation during Lent.

Gospel

- Mark 9:30–37 (LM, #134)
- Luke 4:1–4, 14–21 (LM, #69C)
- John 1:1–18 (LM, #16ABC)
- John 3:14–21 (LM, #32B)
- John 14:15–21 (LM, #55A)

Homily/Reflection

After the proclamation, all are seated for the homily or reflection.

Preaching Points

The presider may use the following points to prepare a homily or reflection focused on the rights and responsibilities given to all persons.

- Human freedom comes from God—not from government or any other source. It is the freedom to pursue, achieve, benefit from, and share our God-given potential with the world. That same freedom requires us to help others to do the same and to share our God-given gifts to our fullest. As Christians, we believe that if basic human rights are denied to anyone that is an injustice to all. Our faith calls us to reflect on the gift of our freedom as adopted sons and daughters of God through Jesus Christ as well as the responsibility that we have for ourselves and each other in honoring that freedom. The Scriptures raise this up in many ways, particularly in the Great Commandment: "You shall love the Lord your God with all your heart, and with all your soul, and with all your strength, and with all your mind; and your neighbor as yourself."[1] Jesus is constantly challenging the abuses of the rights and dignity of the people of Israel by some of the religious authorities.

1. Luke 10:27; see also Matthew 22:37.

- Often rights are denied because of race, ethnicity, religious tradition, or gender. Some of these abuses are written into law; many of them are unwritten laws manifested in institutions and structures in our society. We can also see how policies regarding budgeting affect the rights of others. For example, many nations have a war economy that dwarfs funding of basic human resources for their population, often having an extreme impact of the poor. We also struggle as a nation with the balance between human rights and the common good, often forsaking—even forgetting—the common good in favor of defending individual rights. Take, for example, the issue of marriage. For Catholic Christians, sacramental Marriage is a sacred bond between a couple and God and the community. A sacramental Marriage is always called to be life-giving between and beyond the couple. The individual rights argument often applies only to the specific desires of two individuals, disregarding any greater commitment that their love will be life-giving beyond the two of them. The responsibility in a commitment of consecrated love lies in its unselfish desire to give life to the world beyond the two individuals.

Examination of Conscience

After the homily or reflection, the examination of conscience occurs. Provide the assembly with a time of silence to personally reflect on each examen. The assembly may kneel or stand. The examen should be read aloud by a reader.

Presider: Jesus tells us: "You shall love the Lord your God with all your mind, all your heart, with all your soul, and with all your strength; and your neighbor as yourself."[2] May we ever be open to a deeper conversion within, that we may better use the gifts you have given us, honoring our responsibility to serve those of our brothers and sisters who have been deprived of their basic needs and rights.

Pause for a moment of silent prayer.

Reader: Have I done my very best to support and provide for the basic needs of my family, of my spouse and children?

Pause for a moment of silent prayer.

2. Luke 10:27.

Reader: Do I take the time to support the basic rights and needs of others? Have I been apathetic in defending the rights of people of color, of different ethnicities or religious beliefs?

Pause for a moment of silent prayer.

Reader: Have I supported efforts to house the homeless? To provide food for the hungry? To support quality health care for everyone?

Pause for a moment of silent prayer.

Reader: Have I supported quality education, housing, and health care for children? Have I supported parents in meeting the basic needs of their families?

Pause for a moment of silent prayer.

Reader: Have I worked to support women and girls struggling with decisions regarding their pregnancies?

Pause for a moment of silent prayer.

Reader: Have I engaged in the political process to uphold the rights of others, especially the unborn, the poor, the elderly, ex-offenders, the homeless, and others suffering from injustice?

Pause for a moment of silent prayer.

Reader: Have I placed myself above others, not sharing my gifts, my time, talent, and treasure to their fullest to provide to others the basic needs that I take for granted?

Pause for a moment of silent prayer.

Reader: Do I take the time to learn about people in my own community who are denied their basic rights to food, shelter, and health care? Do I get involved and engage with others to respond?

Pause for a moment of silent prayer.

Reader: Do I take the time to learn about the violations of the basic rights of people in communities beyond my own? Do I pursue ways to help address these injustices?

Pause for a moment of silent prayer.

Reader: Given unique gifts by God, have I fulfilled my own responsibility in developing these gifts and sharing them to their fullest?

Pause for a moment of silent prayer.

Act of Repentance

After the examination of conscience, the presider invites all to pray in these or similar words.

Presider: My brothers and sisters, Jesus proclaims Good News to those gathered on the mountainside and to us today, "Blessed are the poor in spirit, for theirs is the kingdom of heaven!" Humbly and confidently let us ask God to renew us in body, mind, and spirit, that we may more and more conform our lives to reflect the passionate love of our Creator for God's poor.

Reader: "Lord, when did we see you hungry and not feed you?" Each of us hungers for respect, for significance, to be noticed and regarded with dignity. We regard as rights the ability to pursue an education, access good health care, have adequate food and shelter, and live in safety. Yet we fail to recognize the many people in our communities and our nation who do not share in the fullness of those rights and who hunger for those rights for themselves and their children. We are often oblivious to the vast majority of others in our world who share in that collective hunger. We fail to see and teach others that our true significance is found in how we respond to those hungers in others as Christ, who is the source of our freedom, would have us do. Lord, we seek forgiveness and long to know the Master's joy:

R: **Have mercy on us!**

Reader: "Lord, when did we see you thirsty and not give you drink?" So many of us take for granted the freedom we have to pursue our God-given potential, to achieve it, and to enjoy and share its benefits. Yet we fail to acknowledge so many in our society and in our world who thirst for the opportunity to experience the freedoms that we do, to be able to move beyond simply trying to survive each day, and to be able to learn, flourish, and add value to their families and their communities. We fail to engage, educate, and challenge others in our community to fulfill their responsibility in sharing the fullness of their gifts so that others might drink of the freedoms that we take for granted. Lord, we seek forgiveness and long to know the Master's joy:

R: **Have mercy on us!**

Reader: "When did we see you a stranger and not welcome you?" We may often get caught up in cliques in our churches and communities, and we all too often have unwritten rules as to who belongs and who is not welcome. We participate in this as a nation that in practice excludes immigrants and refugees, a nation that deports individuals and separates families. We fail to address the divisions in our own community, the unwritten rules that create insiders and outsiders, friends and strangers. We fail to challenge our communities to be open and regard every human being as a brother and sister in the Lord and to work to ensure that all are welcome and regarded with the same rights and expectations of sharing in the responsibilities of community. Lord, we seek forgiveness and long to know the Master's joy:

R: **Have mercy on us!**

Reader: "When did we see you naked and not clothe you?" We fail to see the effects of rights stripped from people of different races and different faiths, from those who are mentally ill or have served time in prison, or from women who have suffered discrimination and sexual harassment in the workplace. We fail to see the abuses that are inflicted by people in power and to acknowledge our own complicity in not challenging these acts. We fail to educate ourselves and our community on how we can and should respond. We fail to call ourselves and upon our community to a substantive response to help clothe these people in the dignity and rights they deserve. Lord, we seek forgiveness and long to know the Master's joy:

R: **Have mercy on us!**

Reader: "When did we see you ill and not minister to you?" We fail to see the sickness that may inflict us and many of our institutions: the white illness of racism; the male illness of sexism; the sickness of disempowerment; and the ailments of those affected by these forces. We fail to challenge the forces of government, corporations, and our own Church to address this illness and to help bring healing, wholeness, and equality into our society. We fail to see the sickness within our institution of law enforcement and work for the internal cultural change that needs to happen to guarantee that everyone will be treated fairly. Lord, we seek forgiveness and long to know the Master's joy:

R: **Have mercy on us!**

Reader: "When did we see you in prison and not visit you?" We fail to recognize that many of those who are incarcerated are people of color, and the majority of those would not be in prison if they had had the legal representation that most white individuals have. We fail to understand that even upon release, many of these men and women continue to be imprisoned by their prison records, some unable to pursue their right to a job, adequate housing, and, in some states, the ability to vote. We fail to see that we and many of our own people are imprisoned by ignorance of these issues and how to respond. Lord, we seek forgiveness and long to know the Master's joy:

R: **Have mercy on us!**

Reader: "Lord, when did we not see you in our brothers and sisters?" We struggle at times to acknowledge the imprisoned Christ, to leave him behind the bars, separated from our consciousness, freeing us from a need to respond. We fail to recognize those who are imprisoned behind bars, or imprisoned by the inability to rejoin society as responsible citizens because of the discrimination society holds against them. We fail to break down the bars of ignorance in our own community to the plight of those in prison, especially the mentally ill, and to those who are incarcerated because they do not have the resources for a proper defense. Lord, we seek forgiveness and long to know the Master's joy:[3]

R: **Have mercy on us!**

If the assembly has been kneeling for the examination of conscience and act of repentance, they may stand at this time for the remainder of the service.

Presider: By Baptism we are united with Christ and with one another. With one voice, let us pray in the words our Savior gave us.

Our Father . . .

3. The act of repentance is based on Matthew 25:44.

Presider: Lord Jesus, you manifest in us
the call to service of family, neighbor and community,
that all may have the right to freely pursue
their God-given potential.
Open our minds and hearts to recognize more and more
that in the many ways we have been gifted by you,
we are called even more to service of the
freedom of others.
Who live and reign with God the Father in the unity
of the Holy Spirit,
one God, for ever and ever.

All: **Amen.**

Presider: The Lord be with you.

All: **And with your spirit.**

Presider: May almighty God bless you,
the Father, and the Son, and the Holy Spirit.

All: **Amen.**

If a lay minister is the presider, he or she uses this final blessing:

Presider: May the Lord bless us,
protect us from all evil,
and bring us to everlasting life.

All: **Amen.**

Presider: Go in peace, glorifying the Lord by your life.

All: **Thanks be to God.**[4]

4. The greeting, responses, and final dismissal are based in part on those found in *The Roman Missal*, *The Liturgy of the Hours*, and the *Book of Blessings*, International Commission on English in the Liturgy.

Song of Mission

The service concludes with a rousing song commissioning the assembly to go forth in peace and love. Other more familiar songs may also be chosen.

- "Bring Forth the Kingdom" by Marty Haugen (GIA)
- "Lord, Whose Love Through Humble Service" (traditional; various publishers)
- "Onward to the Kingdom" by David Haas (GIA)
- "As a Fire Is Meant for Burning" by Ruth Duck (GIA)
- "You Have Anointed Me" by Gary Daigle and Darryl Ducote (Damean Music/GIA)

THEME 4

The Option for the Poor and Vulnerable

Opening Song

A song from the following options may be chosen or you may select another song that is more familiar to your community.

- "Blest Are They" by David Haas (GIA)
- "The Cry of the Poor" by John Foley, SJ (New Dawn Music)
- "Lord God, Your Love Has Called Us Here" by Brian Wren (Hope Publishing Co.)
- "Christ Be Our Light" by Bernadette Farrell (OCP)
- "The Harvest of Justice" by David Haas (GIA)

After the opening song, the presider and the faithful sign themselves with the Cross. The minister then offers the greeting.

Presider: In the name of the Father, and of the Son, and of the Holy Spirit.

All: **Amen.**

Presider: The Lord be with you.

All: **And with your spirit.**

If the presider is a lay minister, use the following greeting and response:

Presider: Praise the Lord Jesus Christ, who dwells among us. Blessed be God for ever.

All: **Blessed be God for ever.**

Presider: My brothers and sisters, there are times in which we have neglected the poor and the vulnerable. Let us ask God to renew his grace within us as we turn to him in repentance, seeking to change our actions and turning toward those who are in need.

Presider: Let us kneel (**or:** Bow your heads before God).

All pray in silence for a brief period.

Presider: Let us stand (**or:** Raise your heads).

Presider: God of humility and love,
you allowed your Son Jesus to be born into poverty,
that we might know that all is redeemable.
Open our hearts and create a greater space within us,
that we may conform our lives in humble service
to those who are most in need
Through Christ our Lord.

All: **Amen.**

All are seated for the Liturgy of the Word. The following readings with corresponding Lectionary numbers are provided as options. Other readings may also be chosen. Parish readers and cantors may be involved.

First Reading

- Exodus 22:20–26 (LM, #148A)
- Isaiah 61:1–2a, 10–11 (LM, #8B)
- Isaiah 58:1–9a (LM, #231; Weekday)
- Micah 6:1–4, 6–8 (LM, #395. Weekday, Year II)
- 2 Corinthians 8:7, 9, 13–15 (LM, #98B)

Responsorial Psalm

Sing a setting of one of the following psalms that is familiar to your parish.

- Psalm 22:23–24, 26–27, 28, and 31–32
(LM, #924.1; Masses for Various Needs and Occasions)
- Psalm 49:2–3, 6–7, 8–10, 11, 17–18
(LM, #904.2; Masses for Various Needs and Occasions)

- Psalm 107:2–3, 4–5, 6–7, 8–9
  (LM, #924.2; Masses for Various Needs and Occasions)
- Psalm 107:33–34, 35–36, 41–42
  (LM, #929.2; Masses for Various Needs and Occasions)
- Psalm 112:1–2, 3–4, 5–6, 7–8, 9
  (LM, #924.3; Masses for Various Needs and Occasions)

Gospel Acclamation

All stand for the Gospel acclamation and proclamation of the Gospel. Sing a setting of the acclamation that is familiar to your parish. Use the alternate acclamation during Lent.

Gospel

- Matthew 25:31–46 (LM, #160A)
- Luke 4:16–21 (LM, #69C)
- Luke 10:25–37 (LM, #105C)
- Luke 14:1, 7–14 (LM, #126C)
- Luke 16:19–31 (LM, #138C)

After the proclamation, all are seated for the homily or reflection.

Preaching Points

The presider may use the following points to prepare a homily or reflection focused on the preferential option for the poor.

- The irrefutable evidence that God loves the poor is found in the birth of Jesus. Born into the worst of conditions—in a stable, lying in a feeding trough among animals—God's Incarnation among us was an immediate sign that if God could enter into such a dirty, disgusting, smelly place, his redemptive action would seek to touch absolutely everyone. The passionate love of God has a preferential (but not exclusive) regard for the poor.
- The Old Testament texts, especially the prophets, call Israel to task for two major sins against God: the worship of false gods and the neglect of the widow, the orphan, the poor, and the forgotten. The Exodus story regards the enslaved Hebrew people as favored by God and deserving of a Promised Land where they could worship, work, and raise their families free of fear and oppression. The Books of Psalms, Proverbs, and Lamentations carry strong messages of hope

for the poor and the downtrodden and pass severe judgment on those who abuse them.

- God calls us to respond to the needs of the poor who suffer in body, mind, and spirit. At a minimum, we are called to hold them in prayer. And, in prayer, we seek God's wisdom and strength to help each of us discern how God is asking us to respond—to act!—toward the least of our sisters and brothers who are greatest in the eyes of God.

Examination of Conscience

After the homily or reflection, the examination of conscience occurs. Provide the assembly with a time of silence to personally reflect on each examen. The assembly may kneel or stand. The examen should be read aloud by a reader.

Presider: Jesus proclaims to the people of his hometown and to us: "The Spirit of the Lord is upon me, because he has anointed me to bring good news to the poor. He has sent me to proclaim release to the captives and recovery of sight to the blind, to let the oppressed go free, to proclaim the year of the Lord's favor. . . . [T]his scripture has been fulfilled in your hearing!"[1]

Reader: Am I aware of the poor and suffering in my own neighborhood and community? Do I participate in efforts to alleviate that poverty and suffering?

Pause for a moment of silent prayer.

Reader: Have I reflected on my own blindness to the poor and vulnerable in our world? Have I sought clarity in prayer and clear direction from the Lord as to how to respond with the gifts God has given me?

Pause for a moment of silent prayer.

Reader: Have I turned my back on someone in need? Have I made judgments and turned away, even though it may have been Christ himself asking for my kindness?

Pause for a moment of silent prayer.

1. Luke 4:18–19, 21.

Reader: Do I hold stereotypes regarding the poor? Against mothers on welfare, ex-offenders, the addicted, the homeless, or others?

Pause for a moment of silent prayer.

Reader: Do I have a limited sense of the poor? Does my understanding of poverty include everyone who is unable to pursue and develop their God-given potential?

Pause for a moment of silent prayer.

Reader: Am I overwhelmed at times by the poverty and suffering in our world, immobilized by its immensity? Do I seek out the help of God in overcoming this paralysis and moving toward action in the way God may be calling me?

Pause for a moment of silent prayer.

Reader: Do I place the needs of others above my own? Or, do I selfishly cling to material things or to my time rather than share them with those in greater need?

Pause for a moment of silent prayer.

Reader: Do I engage with others in responding to the needs of children, the poor, and the marginalized? Do I seek out organizations that respond to these needs and bring my gifts to assist in responding?

Pause for a moment of silent prayer.

Reader: Have I participated fully in the electoral process, not only voting, but working with others to hold accountable elected officials whose decisions can positively or negatively impact the spiritual, emotional, and material well-being of the poor?

Pause for a moment of silent prayer.

Reader: When my church appeals to help the poor, those affected by war, famine, and natural disasters, do I respond to the best of my ability?

Pause for a moment of silent prayer.

Reader: Do I take time each day to pray for the poor and suffering in our world, being open to where God may be asking me to respond?

Pause for a moment of silent prayer.

Act of Repentance

After the examination of conscience, the presider invites all to pray in these or similar words.

Presider: From the Cross, the Lord says, "Father, forgive them; for do not know what they are doing!"[2] May God forgive us for our failings from inaction or resistance or refusal to provide solace to God's beloved poor.

Reader: "Lord, when did we see you hungry and not feed you?" "Not in my back yard," we often hear from those resistant to soup kitchens and homeless shelters. We prefer not to allow "those people" into the backyards of our consciousness. We fail to acknowledge their hunger and the real hungers of those suffering in the world. We fail to work with others to transform national and international policy so that one day soon no one would need to go hungry; we fail to work with others to save and respect the environment, to conserve energy, to limit deforestation, and to maximize the potential that future generations will be fed. We fail to guarantee the rights of workers to be able to provide basic sustenance for themselves and their families. Lord, we seek forgiveness and long to know the Master's joy:

R: Have mercy on us!

Reader: "Lord, when did we see you thirsty and not give you drink?" We fail to work and support others to help dig wells, to hold governments and corporations accountable for removing pollutants from all water supplies and guaranteeing the free flow of rivers wherever there is need. We fail to work to help those thirsting for justice, for freedom, for peace to achieve

2. Luke 23:34.

those goals through nonviolent means. Lord, we seek forgiveness and long to know the Master's joy:

R: **Have mercy on us!**

Reader: "When did we see you a stranger and not welcome you?" We fail to work with others to welcome orphans into nurturing families, to welcome immigrants searching for a home, and to help change laws so they can have a realistic path toward becoming citizens. We fail to make sure the homeless of our own community and city are cared for, trained for jobs, and given the opportunity to obtain decent and affordable housing. We fail to work with others to create resources for women struggling with their ability to carry a child to term. We fail to work to break down the walls created by racism, sexism, militarism, Islamophobia, homophobia, and arrogant nationalism. Lord, we seek forgiveness and long to know the Master's joy:

R: **Have mercy on us!**

Reader: "When did we see you naked and not clothe you?" We fail to work with others to make sure everyone, no matter what race, ethnicity, or economic class, has a chance to move beyond the nakedness of ignorance and to realize their God-given potential. We fail to help others use their God-given gifts within society through quality education and job training. We fail to remove the nakedness of ignorance from those caught up in sins that create boundaries and limit human potential and creativity. Lord, we seek forgiveness and long to know the Master's joy:

R: **Have mercy on us!**

Reader: "When did we see you ill and not minister to you?" We fail to work with others to create a health system that provides for preventive care and guarantees respectful and thorough health care for all peoples. We fail to support efforts to find a place for the mentally ill who are imprisoned or who live in our communities without receiving proper care and deserved dignity. Lord, we seek forgiveness and long to know the Master's joy:

R: **Have mercy on us!**

Reader: "When did we see you in prison and not visit you?" We fail to work with others to organize prison reform and rehabilitate prisoners. We fail to set up systems to reintegrate nonviolent ex-offenders into society. We fail to work with others within schools to create a nonviolent atmosphere of conflict resolution and to no longer tolerate drug, alcohol, and human abuse. We fail to free people from the prison of addiction, alcoholism, and mental illness. Lord, we seek forgiveness and long to know the Master's joy:

R: **Have mercy on us!**

Reader: "Lord, when did we not see you in our brothers and sisters?" We struggle at times to recognize that the poor are loved in a special way by God and that they deserve our prayers, support, and action. We fail to see Christ in those who beg on street corners, huddle in blankets under viaducts, act strangely because of trauma, or suffer from mental illness. We fail to see and mourn for the Christ dwelling in the casualties of unjust war, genocide, and famine, and the tragic casualties of abortion for both child and mother. Lord, we seek forgiveness and long to know the Master's joy:[3]

R: **Have mercy on us!**

If the assembly has been kneeling for the examination of conscience and act of repentance, they may stand at this time for the remainder of the service.

Presider: When our Lord was born, he had no place to lay his head. He was homeless and born in poverty, thus uniting his sufferings to and sanctifying the plight of humanity. Let us pray in the words he gave us.

Our Father . . .

3. The act of repentance is based on Matthew 25:44.

Presider: Lord Jesus,
How we long to hear those words spoken to us:
"Enter into the joy of your master!"[4]
Make us ever mindful of your call to serve
the least regarded and most vulnerable among us.
Give us the vigilance to notice those you call us to serve
and the courage to risk the costs of real sacrificial love
for the poorest among us.
Who live and reign with God the Father in the unity of the Holy Spirit,
one God, for ever and ever.

All: **Amen.**

Presider: The Lord be with you.

All: **And with your spirit.**

Presider: May almighty God bless you,
the Father, and the Son, and the Holy Spirit.

All: **Amen.**

If a lay minister is the presider, he or she uses this final blessing:

Presider: May the Lord bless us,
protect us from all evil,
and bring us to everlasting life.

All: **Amen.**

Presider: Go in peace, glorifying the Lord by your life.

All: **Thanks be to God.**[5]

4. Matthew 25:23.

5. The greeting, responses, and final dismissal are based in part on those found in *The Roman Missal*, *The Liturgy of the Hours*, and the *Book of Blessings*, International Commission on English in the Liturgy.

Song of Mission

The service concludes with a rousing song commissioning the assembly to go forth in peace and love. Other more familiar songs may also be chosen.

- "Lord, Whose Love In Humble Service" (traditional; various publishers)
- "We Are Called" by David Haas (GIA)
- "We Are Marching" (African traditional; GIA)
- "Send Down the Fire" by Marty Haugen (GIA)
- "Here I Am, Lord" by Daniel L. Schutte (New Dawn Music)

THEME 5

The Dignity of Work and the Rights of Workers

All stand and sing the opening song.

Opening Song

A song from the following options may be chosen or you may select another song that is more familiar to your community.

- "Here Am I" by Brian Wren (GIA)
- "Send Down the Fire" by Marty Haugen (GIA)
- "Lord of All Hopefulness" (traditional)
- "City of God" by Daniel L. Schutte (New Dawn Music)
- "Eye Has Not Seen" by Marty Haugen (GIA)

After the opening song, the presider and the faithful sign themselves with the Cross. The minister then offers the greeting.

Presider: In the name of the Father, and of the Son, and of the Holy Spirit.

All: **Amen.**

Presider: The Lord be with you.

All: **And with your spirit.**

If the presider is a lay minister, use the following greeting and response:

Presider: Let us praise our Lord Jesus Christ, who loved us and gave himself for us. Let us bless him now and for ever. Blessed be God for ever.

All: **Blessed be God for ever.**

After an appropriate song and the greeting by the minister, the meaning of this penitential service is explained to the people. The minister then leads this prayer, which helps the Christian faithful call to mind those times they did not respect the dignity of work and the rights of workers.

Presider: My brothers and sisters, there are times in which we have neglected the dignity of our work and the fundamental rights of workers. Let us ask God to renew his grace within us as we turn to him in repentance, seeking to change our actions and respecting and loving all peoples in this world.

Presider: Let us kneel (**or:** Bow your heads before God).

All pray in silence for a brief period.

Presider: Let us stand (**or:** Raise your heads).

Presider: God of all creation,
in the beginning you formed the earth,
and from its clay shaped woman and man,
breathing into them your Divine Spirit.
May our lives reflect not only the Divine Spirit
that is uniquely within us
but always reverence the dignity and uphold
the right of every person
to share in the ongoing work and the benefits of creation
with you.
Through Christ our Lord.

All: **Amen.**

All are seated for the Liturgy of the Word. The following readings with corresponding Lectionary numbers are provided as options. Other readings may also be chosen. Parish readers and cantors may be involved.

First Reading

- Genesis 1:26—2:3 (LM, #559; Proper of Saints)
- Isaiah 58:1–9a (LM, #221; Weekday)
- Proverbs 8:22–31 (LM, #166C)
- 1 Thessalonians 1:1–5b (LM, #145A)
- James 5:1–6 (LM, #137B)

Responsorial Psalm

Sing a setting of one of the following psalms that is familiar to your parish.

- Psalm 46:2–3, 5–6, 8–9 (LM, #245; Weekday)
- Psalm 90:2, 3–5a, 12–13, 14, 16 (LM, #909.1; Masses for Various Needs and Occasions)
- Psalm 103:1–2, 3–4, 6–7, 8 and 10 (LM, #997.6; Votive Masses)
- Psalm 145:2–3, 4–5, 6–7, 8–9, 10–11 (LM, #945.4; Masses for Various Needs and Occasions)
- Psalm 145:10–11, 15–16, 17–18 (LM, #978.7; Votive Masses)

Gospel Acclamation

All stand for the Gospel acclamation and the proclamation of the Gospel. Sing a setting of the acclamation that is familiar to your parish. Use the alternate acclamation during Lent.

Gospel

- Matthew 11:25–30 (LM, #100A)
- Matthew 20:1–16a (LM, #133A)
- Mark 2:23–28 (LM, #312; Weekday)
- Luke 3:10–18 (LM, #9C)
- Luke 12:13–21 (LM, #114C)

Homily/Reflection

After the proclamation, all are seated for the homily or reflection.

Preaching Points

The presider may use the following points to prepare a homily or reflection that is focused on the dignity of work and the rights of workers.

- On the optional Memorial of St. Joseph the Worker, Pope Francis said in a homily: "We do not get dignity from power or money or culture We get dignity from work. Work is fundamental to the dignity of the person. Work, to use an image, 'anoints' with dignity, fills us with dignity, makes us similar to God who has worked and still works, who always acts."[1] The first papal encyclical on social justice,

1. A Homily given on May 1, 2013.

Rerum novarum,[2] addressed the increased dehumanizing of workers, including child laborers, with the growth of the industrial revolution. Every pope since Leo XIII has made statements affirming the dignity of labor and the rights of workers, including the right to safe working conditions, a livable wage, and the right to organize.

- The Scriptures support the basic dignity of work. "In the beginning," God works for six days to create the universe and the planet on which we live. We believe that God is always at work, not interfering in human free will, but working to build the Kingdom by empowering us to participate as cocreators in God's work. Sadly, individual and corporate greed continue to tear away at the human dignity of workers. Discrimination in the workplace is prevalent: poor working conditions, wage theft (not paying minimum wage, overtime, sick pay, maternity leave, stealing tips, or employers demanding they work a few hours "off the clock" in order to keep their jobs or be reported to immigration authorities). For example, in Chicago, Illinois, wage theft is estimated to be $1 million a day. A study of low-wage workers in large cities found that two-thirds of these workers have been affected by some form of wage theft.[3]
- Although work is an essential component in the lives of our people, we do not often ask our parishioners about their working conditions or whether they have been victims of wage theft, availing them of information about resources, such as, Workers' Centers who can provide pro-bono assistance in recouping lost wages. We also need to support job creation and job training for all who desire to work, as well as seek to remove obstacles such as the stigma against ex-offenders as they seek to make a new life for themselves and their families. We need to support policies that penalize worker abuses and support family friendly policies in the workplace.

---

2. Pope Leo XIII; May 15, 1891.
3. See UIC/UCLA Study, 2008.

Examination of Conscience

After the homily or reflection, the examination of conscience occurs. Provide the assembly with a time of silence to personally reflect on each examen. The assembly may kneel or stand. The examen should be read aloud by a reader.

Presider: The psalmist proclaims: "Let the favor of the Lord our God be upon us, and prosper for us the work of our hands—O prosper the work of our hands!"[4]

Reader: Do I value the job that I have? Do I do my best in the work that I do?

Pause for a moment of silence.

Reader: Have I mistreated my coworkers?

Pause for a moment of silence.

Reader: Have I stood in solidarity with a coworker when they were being treated unfairly?

Pause for a moment of silence.

Reader: Have I treated my employees fairly, offering them a living wage to support their families and creating a safe environment that respects their dignity and encourages their productivity?

Pause for a moment of silence.

Reader: Have I worked to support others in finding meaningful employment in order for them to support their families?

Pause for a moment of silence.

Reader: Have I ever taken advantage of an employee or coworker for my own selfish interests?

Pause for a moment of silence.

Reader: When working conditions have been unsafe, discriminatory, and abusive, have I worked with others to challenge my employer to create a working environment that respects the dignity of all workers?

Pause for a moment of silence.

---

4. Psalm 90:17.

Reader: In overseeing the responsibilities of my job, am I respectful of those above me and those who report to me?

Pause for a moment of silence.

Reader: Have I participated in the political process to ensure the creation of jobs that pay livable wages, offer essential health benefits, and promote a safe working environment?

Pause for a moment of silence.

Reader: Have I encouraged and supported others in pursuing their dreams to start their own business?

Pause for a moment of silence.

Act of Repentance

After the examination of conscience, the presider invites all to pray in these or similar words.

Presider: The psalmist cries out: "I praise you, for I am fearfully and wonderfully made."[5] May our expression of sorrow express our sincerity in bringing hope and healing to all who suffer the indignity of abuse in the workplace and the indignity of being able to provide for themselves and their family.

Reader: "Lord, when did we see you hungry and not feed you?" There are many people who hunger for work, to claim their dignity as human beings and as providers for themselves and their families. Yet we ignore their plight and fail to use our gifts and creativity to help create jobs. We more easily judge them as lazy or unworthy, or fail to notice them at all. We fail to remember that the opposite of love is not hate but apathy and blindness to these members of the human family. Lord, we seek forgiveness and long to know the Master's joy:

R: Have mercy on us!

5. Psalm 139:14.

Reader: "Lord, when did we see you thirsty and not give you drink?" We fail to recognize that all too many who do work thirst for a living wage, for basic benefits, and for safe working conditions. We fail to address the wage theft so prevalent in our own society and challenge employers, even those who claim the name Catholic, to treat their employees with the fairness and dignity they thirst for. Lord, we seek forgiveness and long to know the Master's joy:

R: Have mercy on us!

Reader: "When did we see you a stranger and not welcome you?" Some of us turn our backs on immigrants and refugees, supporting policies that separate families and that send our brothers and sisters back to uncertain futures and sometimes certain death. We fail to recognize that the immigrant adds value to our workforce, to our economy, and to our communities. Instead, we force them to live in the shadows of our society. We fail to hold our elected officials accountable for the psychological, spiritual, and physical pain that our current policies impose on these beloved children of God. Lord, we seek forgiveness and long to know the Master's joy:

R: Have mercy on us!

Reader: "When did we see you naked and not clothe you?" We fail to recognize that immigrants, refugees, and those who have served time in prison are stripped of their God-given freedom to pursue meaningful work, are stripped of the opportunity to claim their dignity, and are stripped of emotional and financial stability to support their families. We fail to challenge government policies that strip workers of the right to organize. We fail to reject the unwritten union policies that prevent equal opportunities for women, immigrants, and people of color. Lord, we seek forgiveness and long to know the Master's joy:

R: Have mercy on us!

Reader: "When did we see you ill and not minister to you?" There is an illness that pervades our society, an illness that blinds us to the plight of the unemployed and the working poor. We fail to recognize the impact this blindness has on families and on home life. We fail to recognize and seek to provide a healing balm on those who are wounded and scarred by unsafe and discriminatory work practices. We fail to challenge employers and government policies that allow unsafe working conditions and ongoing abuse and sexual harassment. Lord, we seek forgiveness and long to know the Master's joy:

R: **Have mercy on us!**

Reader: "When did we see you in prison and not visit you?" We fail to provide opportunities and second chances to those who have served their time. We fail to demand that life skills and job skills training are provided in jails and prisons. We fail to build resources and provide outreach to help these returning citizens find meaningful work. Lord, we seek forgiveness and long to know the Master's joy:

R: **Have mercy on us!**

Reader: "Lord, when did we not see you in our brothers and sisters?" We are often caught up in our own little worlds. We fail to acknowledge that so many among us suffer from prejudicial and retributive attitudes—immigrants, refugees, people of color, women, those of a different sexual orientation. They are the imprisoned Christ, barred from full participation in society. We turn our backs on the Christ who dwells in those who suffer and long for a meaningful job, who simply desire the opportunity to pursue, achieve, and share their God-given gifts. Lord, we seek forgiveness and long to know the Master's joy:[6]

R: **Have mercy on us!**

If the assembly has been kneeling for the examination of conscience and act of repentance, they may stand at this time for the remainder of the service.

6. The act of repentance is based on Matthew 25:44.

Presider: Jesus' foster father, Joseph, was a carpenter, and from Joseph, Jesus learned the value of work. His witness to work fills us with joy. Let us now pray in the words he gave us.

Our Father . . .

Presider: Creator God,
with great love you crafted a world
filled with incredible beauty and diversity.
May that same spirit that breathed on the waters
and breathed into the first man and woman
now fill us with renewed resolve to be faithful
in sharing the fullness of our gifts as cocreators
and coworkers in the building of your Kingdom.
Through Christ our Lord.

All: Amen.

Presider: The Lord be with you.

All: And with your spirit.

Presider: May almighty God bless you,
the Father, and the Son, and the Holy Spirit.

All: Amen.

If a lay minister is the presider, he or she uses this final blessing:

Presider: May the Lord bless us,
protect us from all evil,
and bring us to everlasting life.

All: Amen.

Presider: Go in peace, glorifying the Lord by your life.

All: Thanks be to God.[7]

7. The greeting, responses, and final dismissal are based in part on those found in *The Roman Missal*, *The Liturgy of the Hours*, and the *Book of Blessings*, International Commission on English in the Liturgy.

Song of Mission

The service concludes with a rousing song commissioning the assembly to go forth in peace and love.

- "We Are Called" by David Haas (GIA)
- "We Are Marching" (African American spiritual; GIA)
- "Bring Forth the Kingdom" by Marty Haugen (GIA)
- "We Will Serve the Lord" by Rory Cooney (North American Liturgy Resources/OCP)
- "Lord, Whose Love Through Humble Service" (traditional; various publishers)

THEME 6
Solidarity

All stand and sing the opening song.

Opening Song

A song from the following options may be chosen or you may select another song that is more familiar to your community.

- "Canticle of the Turning" by Rory Cooney (GIA)
- "We Are Many Parts" by Marty Haugen (GIA)
- "Christ Be Our Light" by Bernadette Farrell (OCP)
- "All Are Welcome" by Marty Haugen (GIA)
- "They'll Know We Are Christians" (traditional; various publishers)

After the opening song, the presider and the faithful sign themselves with the Cross. The minister then offers the greeting.

Presider: In the name of the Father, and of the Son, and of the Holy Spirit.

All: **Amen.**

Presider: The Lord be with you.

All: **And with your spirit.**

If the presider is a layperson, use the following greeting and response:

Presider: Let us praise our Lord Jesus Christ, who loved us and gave himself for us. Let us bless him now and for ever. Blessed be God for ever.

All: **Blessed be God for ever.**

After an appropriate song and greeting by the minister, the meaning of this penitential service is explained to the people. The minister then leads this prayer, which helps the Christian faithful call to mind those times they did not stand in solidarity with the peoples of this world.

Presider: My brothers and sisters, there are times in which we have not looked out for others. Let us ask God to renew his grace within us as we turn to him in repentance, seeking to change our actions and stand in solidarity with all peoples in this world who hunger for love, peace, and justice.

Presider: Let us kneel (**or:** Bow your heads before God).

All pray in silence for a brief period.

Presider: Let us stand (**or:** Raise your heads).

Presider: God of unity and peace,
through the life, death, and Resurrection
of your Son Jesus
you have made us your adopted sons and daughter,
all one family held within the life-giving love
of the Most Holy Trinity.
Through this incredible and incomprehensible relationship,
help us to do all in our power to bind
ourselves to one another
by binding our hearts and minds ever more to you.
Through Christ our Lord.

All: **Amen.**

All are seated for the Liturgy of the Word. The following readings with corresponding Lectionary numbers are provided as options. Other readings may be chosen. Parish readers and cantors may be involved.

First Reading

- Romans 13:8–10 (LM, #127A)
- 1 Corinthians 12:12–14, 27–31a (LM, #444; Weekday, Year II)
- Ephesians 4:1–13 (LM, #58B)
- Colossians 3:12–17 (LM, #17ABC)

Responsorial Psalm

Sing a setting of one of the following psalms and canticles that is familiar to your parish.

- Psalm 34:2–3, 4–5, 6–7, 8–9, 10–11 (LM, #978.2; Votive Masses)
- Psalm 40:2–4, 7–8, 8–9, 10 (LM, #64A)
- Psalm 85:2–4, 5–6, 7–8 (LM, #940.2; Masses for Various Needs and Occasions)
- Psalm 113:1–2, 3–4, 5–6, 7–8 (LM, #945.2; Masses for Various Needs and Occasions)
- Isaiah 12:2–3, 4bcd, 5–6 (LM, #985.1; Votive Masses)

Gospel Acclamation

All stand for the Gospel acclamation and the proclamation of the Gospel. Sing a setting of the acclamation that is familiar to your parish. Use the alternate acclamation during Lent.

Gospel

- Matthew 5:1–12a (LM, #70A)
- Matthew 5:20–26 (LM, #228; Weekday)
- Matthew 25:31–46 (LM, #160A)
- John 13:31–33a, 34–35 (LM, #54C)
- John 17:20–26 (LM, #61C)

Homily/Reflection

After the proclamation, all are seated for the homily or reflection.

Preaching Points

The presider may use the following points to prepare a homily or reflection that is focused on human solidarity.

- We are one human family whatever our national, racial, ethnic, economic, and ideological differences. We are our brothers' and sisters' keepers, wherever they may be. We practice solidarity by supporting efforts toward unity that build bridges and break down the artificial and often harmful walls of racism, sexism, nationalism, and xenophobia and other "walls" discriminating against people of different ethnic background, religious belief, social class, or sexual orientation.
- The Scriptures remind us that we all share the spirit of God, that we are all made in God's image and likeness. Throughout the Old and New Testaments we hear how individuals and nations violated God's desire for peoples to live as one human family in peace and harmony. Jesus expressed this in his final discourse in the Gospel according to John: "that they may all be one. As you, Father, are in me and I am in you, may they also be in us, so that the world may believe that you have sent me."[1] If we examine the healing stories of Jesus, they are most often not just about healing that brings wholeness to a person's body but, even more so, bringing them into wholeness within themselves and within the community.
- We can easily identify ways that we violate this principle of solidarity: gossip, jealousy, envy, talking about and talking at others (rather than talking with them), stereotyping groups of people.
- There are larger social forces that threaten solidarity: white nationalism, radical extremism, and public policies that create second-class citizens by applying laws disproportionately against black and Latino populations and continuing to penalize them after serving time in prison.
- Our call as people of faith is to promote and work to achieve solidarity within our homes, within our communities, within our Church and school, and within our nation and world.

---

1. John 17:21.

Examination of Conscience

After the homily, the examination of conscience occurs. Provide the assembly with a time of silence to personally reflect on each examen. The assembly may kneel or stand. The examen should be read aloud by a reader.

Presider: Even at the Cross, Jesus held all of us in mutual care. "Standing near the cross of Jesus were his mother, and his mother's sister, Mary the wife of Clopas, and Mary Magdalene. When Jesus saw his mother and the disciple whom he loved standing beside her, he said to his mother, 'Woman, here is your son.' Then he said to the disciple, 'Here is your mother.' And from that hour the disciple took her into his own home."[2]

Reader: Have I contributed to the well-being of my family? Have I encouraged healthy conversation, respectful dialogue, peaceful resolution of conflict, and shared responsibility in our household?

Pause for a moment of silent prayer.

Reader: Do I hold prejudices toward people of different races, ethnicities, or religious beliefs? Do I participate in conversations that demean certain groups of people because of race, gender, religious belief, or sexual orientation?

Pause for a moment of silent prayer.

Reader: Do I work to prevent violence against others in all of its forms, such as gun violence; domestic violence; verbal, physical, and sexual abuse of children and youth; and racial and ethnic violence?

Pause for a moment of silent prayer.

Reader: Do I support efforts that empower others to pursue and share their God-given gifts?

Pause for a moment of silent prayer.

2. John 19:25–29.

Reader: Do I hold in prayer all those who suffer injustice in our world, all precious children of God?

Pause for a moment of silent prayer.

Reader: Do I participate in activities that support the dignity of my brothers and sisters throughout the world?

Pause for a moment of silent prayer.

Reader: Have I sought out ways to help those whose rights and dignity are being violated? In my own community? In my city? In my country? In the world?

Pause for a moment of silent prayer.

Reader: Do I belong to organizations that support the rights of workers and the rights of all who desire to work?

Pause for a moment of silent prayer.

Reader: Do I seek ways to hold those in decision-making positions in government accountable for the fair treatment of all my fellow citizens?

Pause for a moment of silent prayer.

Reader: Do I work to uphold the rights of immigrants and migrants seeking safety and security, those who seek to affirm their dignity by caring for and supporting their families?

Pause for a moment of silent prayer.

Act of Repentance

After the examination of conscience, the presider invites all to pray in these or similar words.

Presider: The psalmist tells us: "How very good and pleasant it is when kindred live together in unity!"[3] Jesus expresses the deepest longing in his heart, "That they may all be one!"[4] May our expression of sorrow and openness to conversion reflect our desire to honor God's passion that humankind may live in true harmony and lasting peace.

3. Psalm 133:1.
4. John 17:21.

Reader: "Lord, when did we see you hungry and not feed you?" There are many people in our world who feel alone and abandoned, who hunger for belonging and simple recognition and affirmation. Yet we fail to see that some of these people may be part of our own family, our staff, our coworkers, or our congregation. We fail to recognize our own wastefulness while at the same time ignore the real physical, emotional, and spiritual hunger so prevalent in our community. We fail to challenge our elected officials to recognize and respond to the legitimate hungers in every member of our society. Lord, we seek forgiveness and long to know the Master's joy:

R: Have mercy on us!

Reader: "Lord, when did we see you thirsty and not give you drink?" Some say that the next great war will be over fresh water, that communities and nations will be divided by the need for and acquisition of potable water. Yet we fail to recognize that many in our own country thirst for potable water while we often waste the water that we so easily take for granted. We fail to witness the despair and sense of abandonment in those communities that are a part of our own country who feel that the rest of us do not care. We fail to call our elected officials to correct this injustice against our brothers and sisters. Lord, we seek forgiveness and long to know the Master's joy:

R: Have mercy on us!

Reader: "When did we see you a stranger and not welcome you?" We make judgments against the immigrant and the refugee and allow our government to continue to tear families apart, ultimately tearing at the very fabric of the diversity that has defined us and given us pride as a nation. We fail to challenge the forces of division, of white supremacy, of those who seek to exclude as not deserving of full recognition and inclusion in our country. We make judgments against those who would seek to stand or kneel to raise awareness and call for equal treatment of every citizen and noncitizen.

We fail to address our own national policies that tear at the fabric of the human community through an out of control war economy and economic policies that divide rather than seek to unify. Lord, we seek forgiveness and long to know the Master's joy:

R. **Have mercy on us!**

Reader: "When did we see you naked and not clothe you?" We fail to recognize the many people in our society and in our world who are stripped of their dignity, who are labeled as "less important than, less worthy of, or of less value than others." We fail to clothe ourselves with information, education, and ways to respond to these least among us. We fail to raise awareness in our own communities about those who do not have the opportunity for full participation in society because of poor educational institutions, lack of jobs, and livable and affordable housing. We fail to address national and international policies that strip so many of the freedoms and opportunities that we take for granted. Lord, we seek forgiveness and long to know the Master's joy:

R: **Have mercy on us!**

Reader: "When did we see you ill and not minister to you?" There are many illnesses that threaten our solidarity within families, within Church communities, within our local communities, and within our nation. Racism, sexism, discrimination based on religion or sexual orientation, or political ideology are all illnesses that infect many of us and cause so much emotional and spiritual pain to those who are the object of these "–isms." Yet we fail to build bridges, reach out and educate, live out the mandate of our faith to address the divisions within ourselves, to challenge those who seek to divide. We fail to accompany the victims of these "–isms" as they seek healing and restoration. Lord, we seek forgiveness and long to know the Master's joy:

R: **Have mercy on us!**

Reader: "When did we see you in prison and not visit you?" There are so many in our society imprisoned by prejudice and racism. Many in our society are imprisoned by fear, feeling the need to build walls around their communities and have guns to protect themselves. Yet we fail to work to break down these bars with education and action that promote faith in God and faith in the inherent goodness of our brothers and sisters. We fail to challenge government policies designed to isolate and imprison certain populations in fear and uncertainty. We fail to address the prejudices within ourselves that prevent us from taking the risks necessary to help others who are different from us. Lord, we seek forgiveness and long to know the Master's joy:

R: Have mercy on us!

Reader: "Lord, when did we not see you in our brothers and sisters?" We fail to take the time in prayer to include the whole human community in our petitions, to take the time to see the faces of those in the human family who suffer each day. We fail to see what Christ needs and longs for us to see in order that we might participate in bringing healing and wholeness to his body in this world. Lord, we seek forgiveness and long to know the Master's joy:[5]

R: Have mercy on us!

If the assembly has been kneeling for the examination of conscience and act of repentance, they may stand at this time for the remainder of the service.

Presider: Through Baptism, we are one in God and one in the human family. Let us pray in the words his Son gave us.

Our Father . . .

Presider: God of justice and peace,
your Son Jesus reminds us that true peace
is found only in working to ensure peace
    through justice for others.
Bind us to you: our heart with your heart,
    our mind with your mind,

5. The act of repentance is based on Matthew 25:44.

our dream with your dream, that, united with you,
we may work to achieve that longed-for unity
found in the full flourishing
of all humankind.
Through Christ our Lord.

All: **Amen.**

Presider: The Lord be with you.

All: **And with your spirit.**

Presider: May almighty God bless you,
the Father, and the Son, and the Holy Spirit.

All: **Amen.**

If a lay minister is the presider, he or she uses this final blessing:

Presider: May the Lord bless us,
protect us from all evil,
and bring us to everlasting life.

All: **Amen.**

Presider: Go in peace, glorifying the Lord by your life.

All: **Thanks be to God.**[6]

Song of Mission

- "Bring Forth the Kingdom" by Marty Haugen (GIA)
- "World Peace Prayer" by Marty Haugen (GIA)
- "What You Have Done for Me" by Tony Alonso (GIA)
- "Prayer of Peace" by David Haas (GIA)
- "Send Down the Fire" by Marty Haugen (GIA)

6. The greeting, responses, and final dismissal are based in part on those found in *The Roman Missal*, *The Liturgy of the Hours*, and the *Book of Blessings*, International Commission on English in the Liturgy.

THEME 7
Care for God's Creation

All stand and sing the opening song.

Opening Song

A song from the following options may be chosen or you may select another song that is more familiar to your community.

- "All Creatures of Our God and King" (traditional; various publishers)
- "Sing Out Earth and Skies" by Marty Haugen (GIA)
- "Spirit Blowing through Creation" by Marty Haugen (GIA)
- "Shall We Gather at the River" (traditional; various publishers)
- "Joyful, Joyful We Adore You" (traditional; various publishers)

After the opening song, the presider and the faithful sign themselves with the Cross. The minister then offers the greeting.

Presider: In the name of the Father, and of the Son, and of the Holy Spirit.

All: **Amen.**

Presider: The Lord be with you.

All: **And also with you.**

If the presider is a lay minister, use the following greeting and response:

Presider: Praise be our God who made heaven and earth. The sun and the moon, the stars and the sea, the birds, and the land. Most High, all powerful, good Lord, Yours are the praises, the glory, the honor, and all blessing. Blessed be God for ever.[1]

All: **Blessed be God for ever.**

---

1. Adapted from St. Francis' *Canticle of the Sun.*

After an appropriate song and greeting by the minister, the meaning of this penitential service is explained to the people. The minister then leads this prayer, which helps the Christian faithful call to mind those times that they have failed to care for and respect God's creation.

Presider: My brothers and sisters, our earth "now cries out to us because of the harm we have inflicted on her by our irresponsible use and abuse of the goods with which God has endowed her. We have come to see ourselves as her lords and masters, entitled to plunder her at will. The violence present in our hearts, wounded by sin, is also reflected in the symptoms of sickness evident in the soil, in the water, in the air and in all forms of life. This is why the earth herself, burdened and laid waste, is among the most abandoned and maltreated of our poor; she 'groans in travail.' We have forgotten that we ourselves are dust of the earth, our very bodies are made up of her elements, we breathe her air and we receive life and refreshment from her waters."[2] Let us ask God to renew his grace within us, as we turn to him in repentance, seeking to change our actions, respecting and loving all that lives in this world.

Presider: Let us kneel (**or:** Bow your heads before God).

All pray in silence for a brief period.

Presider: Let us stand (**or:** Raise your heads).

All-powerful God, you are present in the whole universe
and in the smallest of your creatures.
You embrace with your tenderness all that exists.
Pour out upon us the power of your love,
that we may protect life and beauty.
Fill us with peace, that we may live
as brothers and sisters, harming no one.
O God of the poor,
help us to rescue the abandoned and forgotten
of this earth,
so precious in your eyes.

2. *Laudato si'* (LS), 2.

Bring healing to our lives,
that we may protect the world and not prey on it,
that we may sow beauty, not pollution and destruction.
Touch the hearts
of those who look only for gain
at the expense of the poor and the earth.
Teach us to discover the worth of each thing,
to be filled with awe and contemplation,
to recognize that we are profoundly united
with every creature
as we journey toward your infinite light.
We thank you for being with us each day.
Encourage us, we pray, in our struggle
for justice, love, and peace.
Through Christ our Lord.
Amen.[3]

All are seated for the Liturgy of the Word. The following readings with corresponding Lectionary numbers are provided as options. Other readings may also be chosen. Parish readers and cantors may be involved.

First Reading

- Genesis 1:1—2:2 (LM, #41ABC)
- Deuteronomy 10:12–22 (LM, #413; Weekday, Year I)
- Romans 1:16–25 (LM, #468; Weekday, Year I)
- Acts of the Apostles 17:15, 22—18:1 (LM, #293; Weekday)
- Colossians 1:15–20 (LM, #105C)

Responsorial Psalm

Sing a setting of one of the following psalms that is familiar to your parish.

- Psalm 8:4–5, 6–7, 8–9 (LM, #904.1; Masses for Various Needs and Occasions)
- Psalm 65:10, 11–12, 13–14 (LM, #914.1; Masses for Various Needs and Occasions)
- Psalm 104:1–2a, 14–15, 24, 27–28 (LM, #914.2; Masses for Various Needs and Occasions)

---

3. LS, 246.

- Psalm 107:35–36, 37–38, 41–42 (LM, #914.3; Masses for Various Needs and Occasions)
- Psalm 126:1–2, 2–3, 4–5, 6 (LM, #6C)

Gospel Acclamation

All stand for the Gospel acclamation and the proclamation of the Gospel. Sing a setting of the acclamation that is familiar to your parish. Use the alternate acclamation during Lent.

Gospel

- Matthew 6:24–34 (LM, #82A)
- Mark 4:1–20 (LM, #319; Weekday)
- Mark 4:35–41 (LM, #95B)
- John 1:1–18 (LM, #19ABC)
- John 15:1–8 (LM, #53B)

Homily/Reflection

After the proclamation, all are seated for the homily or reflection.

Preaching Points

The presider may use the following points to prepare a homily or reflection focused on the care for creation.

- All creation is gift! The world in which we live, the air we breathe, the water we drink, all is a gift from God. We give respect to the Creator by our thoughtful stewardship of his creation. Our faith and sense of morality demand that we protect the environment and be considerate toward future generations of people living on this planet.
- The Scriptures are replete with images of creation, of its beauty, and of its destructive force. Throughout his ministry, Jesus used a multitude of images from nature: recall the parable of the mustard seed, references to the lilies of the field, the sparrow, a fig tree, a mother hen protecting her children, the sower and the seed, and many others. The Scriptures also remind us that the environment is not something apart from us but that we are active members who influence the physical, emotional, and spiritual environment of our own bodies and our families.

- As Christians we are called to regard our bodies and our planet as sacred gift. We are asked to defend the environment from forces of greed that are destroying whole species of plants and animals, devastating populations, causing greater harm to those living in impoverished areas of our planet, and putting at great risk the ability for future generations to survive and thrive.
- We are called to act locally where we are able yet always be mindful and open to ways to protect the environment through organizations supporting the environment and advocating against any laws that would cause further harm so that the earth may continue to provide for us and for many generations to come.
- The presider should look to Pope Francis' document, *Laudato si'*, for scientific and theological rationale on protecting the earth and its connection to Christian discipleship.

Examination of Conscience

After the homily or reflection, the examination of conscience occurs. Provide the assembly with a time of silence to personally reflect after each examen. The assembly may kneel or stand. The examen should be read aloud by a reader.

Presider: St. John tells us: "In the beginning was the Word, and the Word was with God, and the Word was God. He was in the beginning with God. All things came into being through him, and without him not one thing came into being. What has come into being in him was life, and the life was the light of all people. The light shines in the darkness, and the darkness did not overcome it."[4]

Reader: Do I see myself as part of the environment, a precious member of God's creation? Do I care for my body, my mind, my spirit, that I might honor the gift of life God has given me?

Pause for a moment of silent prayer.

Reader: Do I nurture the environment of my home? Do I participate in the sharing of responsibilities within the home? Do I encourage my family members in proper and healthy self-care? Is prayer a key component of the environment in my home? Do I encourage an environment of mutual

4. John 1:1–5.

sharing, healthy dialogue, peaceful resolution of conflict, and the sharing of love?

Pause for a moment of silent prayer.

Reader: Have I encouraged my family members to give thanks for the gift of life and the gift of the world God has given us? Have I encouraged and nurtured a spirit of awe at God's creation?

Pause for a moment of silent prayer.

Reader: Have I been wasteful with my food? Do I recycle and encourage others to do the same?

Pause for a moment of silent prayer.

Reader: Have I sought to live a life of simplicity, freeing myself and my environment from distractions and unnecessary clutter in my life? Do I resist the urge to accumulate things, especially the wants that can often be perceived as needs?

Pause for a moment of silent prayer.

Reader: Do I care for the environment outside of my home and in my neighborhood? Do I pick up litter and do my best to beautify the area around my home?

Pause for a moment of silent prayer.

Reader: Do I support efforts that respect the environment through participation in organizations that promote clean energy, the diminishment of pollution, and the preservation of natural resources and the ongoing upkeep and care for our national and state parks?

Pause for a moment of silent prayer.

Reader: Do I participate in the political process by working with others to hold elected officials accountable to support policies that respect the environment?

Pause for a moment of silent prayer.

Reader: Do I work with organizations that challenge corporations and businesses that pollute the environment, often affecting the health and well-being of poor and vulnerable communities?

Pause for a moment of silent prayer.

Reader: Do I reflect each day on the gift of my life and the resources of this earth that sustain me and those I love? Do I ask God for wisdom to respond to the needs of others who suffer due to the abuses of God's creation?

Pause for a moment of silent prayer.

Act of Repentance

After the examination of conscience, the presider invites all to pray in these or similar words.

Presider: "God, who calls us to generous commitment and to give him our all, offers us the light and the strength needed to continue on our way. In the heart of this world, the Lord of life, who loves us so much, is always present. He does not abandon us, he does not leave us alone, for he has united himself definitively to our earth, and his love constantly impels us to find new ways forward."[5] May our expression of sorrow and desire for conversion manifest itself in greater care for all creation.

Pause for a moment of silent prayer.

Reader: "Lord, when did we see you hungry and not feed you?" We are often wasteful of our food and our time. We fail to feed our families with an appreciation of God's gifts, of the basics of life, and teach them that there are so many others in our world who do not have what we have—even the basics of food and drinkable water. We fail to appreciate that we have the opportunity to take a Sabbath each week, a day set aside for worship and community, where many in our world hunger for and yet do not have that opportunity. Lord, we seek forgiveness and long to know the Master's joy:

R: Have mercy on us!

---

5. LS, 245.

Reader: "Lord, when did we see you thirsty and not give you drink?" Our planet thirsts for recognition and respect, and more and more it becomes parched and less habitable for all living things due to global warming caused by our human activity. Yet we take for granted the comfortable environment in which we live, failing to open our eyes, our hearts, and our minds to the ways that our planet and our brothers and sisters are being ravaged by the effects of our excesses and our government policies, even those in our own community and country. We fail to recycle, compost, use minimal water, become less wasteful with our food, and enjoy a simpler lifestyle. Lord, we seek forgiveness and long to know the Master's joy:

R: **Have mercy on us!**

Reader: "When did we see you a stranger and not welcome you?" We are not aware of the high rate of extinction of so many species who inhabit our planet. We do not mourn the loss of life—human, animal, and plant—and seek ways to preserve life in all its manifestations; to us they pass from the earth as a stranger we never encountered or cared about. We fail to work with those who are trying to save our planet, save God's beloved creation, and challenge those in government and business to place life above greed, to regard everything as a precious creation and gift to us all. Lord, we seek forgiveness and long to know the Master's joy:

R: **Have mercy on us!**

Reader: "When did we see you naked and not clothe you?" So much of our planet has been stripped of its dignity and resources. Yet we fail to fight for the life of our planet. We fail to reverence the sacrifices of past generations who passed these resources on to us. We fail to see that our children and future generations will be stripped of resources they will need to simply survive. We fail to challenge political and corporate forces that continue to abuse and strip our planet, the poor, and future generations of resources needed to sustain their lives and allow them to flourish. Lord, we seek forgiveness and long to know the Master's joy:

R: **Have mercy on us!**

Reader: "When did we see you ill and not minister to you?" Our planet is sick. Those who are making decisions negatively affecting the environment are sick with greed and arrogance and total disregard for the poor. Yet we stand by as many people are deprived of basic health care and the medicines they need. We stand by as new viruses ravage poor communities. We fail to educate ourselves and our families about these sins against humanity and the earth and fail to come to their defense. Lord, we seek forgiveness and long to know the Master's joy:

R: **Have mercy on us!**

Reader: "When did we see you in prison and not visit you?" So many people are imprisoned by ignorance regarding the effects of their personal actions and the action or inaction of our government on the environment. They are ignorant of the effects of these actions on the lives of people in our own communities, cities, country, and the world. We fail to tear down the bars and break the chains that hold so many on our planet hostage to the apathy and the negative actions of so many who are imprisoned in their false securities and ignorance. Lord, we seek forgiveness and long to know the Master's joy:

R: **Have mercy on us!**

Reader: "Lord, when did we not see you in our brothers and sisters?" All too often, we fail to see the presence of God in nature and in our fellow human beings. We fail to give witness to the eternal Christ who calls us to build up and not tear down, to be life givers and not life takers. Caught up in our own lives, in technology, and by so many distractions, we fail to be awed by the grandeur of creation and of the Spirit of God alive in us—that same spirit that breathed order into chaos, and soul and spirit into the first human beings.

We fail to call our congregations to recognize that God is indeed love but also the beauty found in the created order in which we dwell. Lord, we seek forgiveness and long to know the Master's joy:[6]

R: Have mercy on us!

If the assembly has been kneeling for the examination of conscience and act of repentance, they may stand at this time for the remainder of the service.

Presider: The destiny of all creation is bound up with the mystery of Christ, present from the beginning. Let us pray in the words he gave us.[7]

Our Father . . .

Presider: God of love,
show us our place in this world as channels of your love
for all the creatures of this earth,
for not one of them is forgotten in your sight.
Enlighten those who possess power and money,
that they may avoid the sin of indifference,
that they may love the common good,
advance the weak,
and care for this world in which we live.
The poor and the earth are crying out.
O Lord, seize us with your power and light,
help us to protect all life,
to prepare for a better future
for the coming of your Kingdom
of justice, peace, love, and beauty.
Praise be to you!
Through Christ our Lord.
Amen.[8]

Presider: The Lord be with you.

All: And with your spirit.

6. The act of repentance is based on Matthew 25:44.
7. Adapted from LS, 99.
8. LS, 246.

Presider: May almighty God bless you,
the Father, and the Son, and the Holy Spirit.

All: **Amen.**

If a lay minister is the presider, he or she uses this final blessing:

Presider: May the Lord bless us,
protect us from all evil,
and bring us to everlasting life.

All: **Amen.**

Presider: Go in peace, glorifying the Lord by your life.

All: **Thanks be to God.**[9]

Song of Mission

The service concludes with a rousing song commissioning the assembly to go forth in peace and love. Another more familiar song may be chosen.

- "For the Beauty of the Earth" (traditional; various publishers)
- "How Great Thou Art" (traditional; various publishers)
- "Find Us Faithful" by Steve Green (GIA)
- "Let All Things Now Living" (traditional; various publishers)
- "Canticle of the Sun" by Marty Haugen (GIA)

9. The greeting, responses, and final dismissal are based in part on those found in *The Roman Missal*, *The Liturgy of the Hours*, and the *Book of Blessings*, International Commission on English in the Liturgy.

ABOUT THE AUTHORS

Kevin Ahern, PhD, is an assistant professor of religious studies at Manhattan College, where he directs the peace studies program. He has published several books, including the award-winning *Visions of Hope: Emerging Theologians and the Future of the Church, The Radical Bible*, and *Structures of Grace: Catholic Organizations Serving the Global Common Good*. He is also a coeditor of *Public Theology and the Global Common Good*. From 2003 to 2007, he served as president of the International Movement of Catholic Students, a network of Catholic students in more than eighty countries. He presently serves as president of the International Catholic Movement for Intellectual and Cultural Affairs (ICMICA–Pax Romana), a global movement of Catholic professionals and intellectuals committed to Catholic social teaching.

Larry Dowling is a priest of the Archdiocese of Chicago. He was ordained in 1991 after receiving an MDIV from Mundelein Seminary. He acquired a DMIN from the University of St. Mary of the Lake in 1997 and is currently the pastor of St. Agatha Catholic Parish in Chicago. He actively serves on many peace and justice committees for both Chicago and the Archdiocese of Chicago. He is a recipient of the Harvest of Hope Award from the Archdiocese of Chicago's immigration office. He has been a reporter for the National Federation of Priests' Councils' Annual Conference for the last twenty years; he served for eighteen years as the editor of UPTURN, a publication of the Association of Chicago Priests; and he has written for LTP's *Pastoral Liturgy* and *Sourcebook for Sundays, Seasons, and Weekdays*. Fr. Dowling enjoys writing, biking, and backpacking.

Bernard Evans, PhD is retired faculty at St. John's University, Collegeville, Minnesota, where he serves as associate dean for faculty in the school of theology. Evans also occupies the Virgil Michel Ecumenical Chair in Rural Social Ministries, teaching courses on Christian social ethics, environmental theology, and ministry in rural communities. His most recent publications include the books *Lazarus at the Table: Catholics and Social Justice* (2006); *Vote Catholic? Beyond the Political Din* (2008); *Stewardship: Living a Biblical Call* (2014); as well as a chapter, "Care for Creation," in *A Vision of Justice*, edited by Ron Pagnucco and Susan Crawford Sullivan (2014), all published by Liturgical Press.

Anne Y. Koester, JD, MA (theology), works at Georgetown University, Washington, DC, where she is also an adjunct instructor with the theology department. In addition, she oversees the RCIA process at Holy Trinity Catholic Church in DC. A former trial lawyer, Anne studied theology at St. John's University in Collegeville, Minnesota. She has worked at the Notre Dame Center for Pastoral Liturgy and the Georgetown Center for Liturgy. Anne is the author of *Sunday Mass: Our Role and Why It Matters* (Liturgical Press, 2007), editor of *Liturgy and Justice: To Worship God in Spirit and in Truth* (Liturgical Press, 2002), and coeditor of *Vision: The Scholarly Contributions of Mark Searle to Liturgical Renewal* (Liturgical Press, 2004) and *Called to Participate: Theological, Ritual and Social Perspectives* by Mark Searle (Liturgical Press, 2006). She is a member of the editorial advisory council of Liturgical Press and the North American Academy of Liturgy.

Thomas Massaro, SJ, is professor of moral theology and former dean at the Jesuit School of Theology of Santa Clara University, a graduate school of ministry located in Berkeley, California. A Jesuit priest of the Northeast Province, he was professor of moral theology for fifteen years at Weston Jesuit School of Theology in Cambridge, Massachusetts, and at Boston College before arriving in Berkeley, California, in 2012. In the summer of 2018 he took up a new teaching appointment in theological ethics at Fordham University in New York City. Father Massaro holds a doctorate in Christian social ethics from Emory University. His several books and many articles are devoted to Catholic social teaching and its recommendations for public policies oriented to social justice, peace, worker rights, and poverty alleviation. A former columnist for *America* magazine, he writes and lectures frequently on such topics as the ethics of globalization, peacemaking, environmental concern, the role of conscience in religious participation in public life, and developing a spirituality of justice. Besides teaching courses on many aspects of Catholic social teaching and the role of religion in public life, he seeks to maintain a commitment to hands-on social activism. He served a six-year term on the Peace Commission of the City of Cambridge and is a cofounder and national steering committee member of Catholic Scholars for Worker Justice.

Dawn M. Nothwehr, OSF, PhD, holds the Erica and Harry John Family Endowed Chair in Catholic Theological Ethics at Catholic Theological Union in Chicago. The mandate of the endowed chair is to promote the Roman Catholic Consistent Ethic of Life, advanced by Cardinal Bernardin. She teaches courses in environmental ethics, racial justice, feminist ethics, and fundamental moral theology and has published many books and articles on these same topics.

Timothy P. O'Malley, PHD, is academic director of the Notre Dame Center for Liturgy and managing director of the McGrath Institute for Church Life. He has a concurrent position in the department of theology at the University of Notre Dame, where he teaches and researches in the area of liturgical-sacramental theology, catechesis, and spirituality. He is the author of *Liturgy and the New Evangelization: Practicing the Art of Self-Giving Love* (Liturgical Press, 2014) and *Bored Again Catholic: How the Mass Could Save Your Life* (Our Sunday Visitor, 2017).

Thomas Scirghi, SJ, is a native of New York City and a member of the Jesuit Order. He is associate professor of theology at Fordham University, where he teaches sacramental and liturgical theology. He is also the rector for the Jesuit Community of Fordham. Previously he taught at the Jesuit School of Theology in Berkeley, California. There he instructed students—seminarians and laity—in the theory and practice of liturgy. Fr. Scirghi is the author of numerous articles and several books, most recently *Longing to See Your Face: Preaching in a Secular Age* (Liturgical Press). He is also the author of "Signs of God's Grace: A Journey through the Sacraments" (Now You Know Media), a series of lectures for CD and DVD. Fr. Scirghi has lectured around the United States, Africa, Nepal, and Australia. He explains that his chief concern for the Roman Catholic Church today is to make our tradition accessible to the next generation.

Kate Ward, PHD, is assistant professor of theological ethics at Marquette University. Her published and forthcoming articles on questions of economic ethics, virtue ethics, and ethical method appear in journals including *Theological Studies, Journal of the Society of Christian Ethics,* and *Journal of Religious Ethics*. She recently coedited an issue of the journal *Religions*, focused on economic inequality, with Kenneth R. Himes, and is at work on a book manuscript titled "Wealth, Virtue, and Moral Luck: Christian Ethics in an Age of Inequality." Dr. Ward graduated in 2005 from Harvard College, where she studied psychology, and she earned her MDIV with concentration in Bible from Catholic Theological Union in 2011. Before beginning her PHD in 2016 from Boston College, she worked at AFSCME Council 31, a labor union organizing workers in Catholic healthcare settings.